The Living Gut

The Living Gut

An Introduction to Micro-organisms in Nutrition

WN Ewing B Sc, PhD, MIBiol.

DJA Cole B Sc, PhD, FIBiol, FRSA
University of Nottingham, England.

A Con*text* Publication.

Con*text*,
117 Carrycastle Road, Dungannon,
Co Tyrone, N. Ireland, BT70 1LT

First published 1994

British Library Cataloguing in Publication Data
The Living Gut
I. Ewing, W.N.
II. Cole, D.J.A.

ISBN 1–899043–00–4

Typesetting by The Midlands Book Typesetting Company, Loughborough, England.
Illustrations and design by Con*text* Graphics, Dungannon, N. Ireland

Printed and bound by Redwood Books, Trowbridge, Wiltshire

CONTENTS

PREFACE

The alimentary tract of mammals is the essential organ which extracts nutrients from feedstuffs and allows their absorption and incorporation into the blood circulation. As such it represents an interface between the metabolism of the animal and the environment. Consequently, environmental influences can play an important part in the complex set of relationships in the gut.

A key feature of the alimentary tract is its ability to digest foodstuffs before absorption. In this, hydrolysis plays an important part. However, its functions are not confined to chemical reactions. Digestion occurs partly through fermentation which, while important in monogastric animals, is even more so in ruminants. Consequently, the microbial population is of particular importance.

This book seeks to present basic information on micro-organisms, particularly bacteria, in the nutrition of the host animal. At the same time, the value of other bacterial functions is recognised. Although commercial development of bacterial products will undoubtedly occur, the essential principles upon which they are based will not alter. Interest in the function of the gut is directed towards, amongst others, man, farm livestock and companion animals.

As far as farm livestock are concerned, the object is to ensure conditions which allow the most efficient use of feedstuffs and particularly the poorer materials, e.g. forages, for which man does not compete. Invariably good gut health means high efficiency.

As far as man is concerned, the situation is different. There is clearly a considerable emphasis on health, but efficient nutrition and the meeting of daily requirements are not the only other criteria. Eating is an emotional experience and when food is in plentiful supply the consuming public are more likely to consider safety, ethical and moral issues, together with demands related to their perception of a wholesome diet.

For example, there is increasing concern about the over-use of chemicals and drugs throughout the world. This concern is fuelled by reports of resistant strains of bacteria associated with diseases previously thought to be nearly irradicated. Tuberculosis had almost disappeared through the widespread use of antibiotics and vaccines after the Second World War. However, in 1990 and 1991 in the United States 13 outbreaks of tuberculosis were reported, caused by a strain of bacteria that was resistant to two or more of the most powerful anti-TB drugs, e.g. isoniazid and rifampacin, (Kingman, 1993). Raviglione from the World Health Organisation stated, "If the bacteria are resistant to all drugs we have available, the disease becomes practically incurable. It becomes the disease it was at the beginning of the century". Man is now looking at alternatives to maintain growth and performance in farm livestock, while at the same time considering the health, safety and acceptability of his own diet.

This book seeks to consider those materials which particularly influence the nature of the microbial balance of the gut. Such materials may be living or dead organisms or chemicals. Eighty years ago Eli Metchinkoff observed that the regular consumption of fermented milk (yoghurt) containing *Lactobacillus bulgaricus* and *Streptococcus thermophilus* was beneficial to human health. The same theory still applies today with the eating of live yoghurts.

There is considerable interest by consumers in many countries, in the nature and safety of their own food. In addition, there is a wealth of advice on healthy eating from such organisations as the World Health Organisation and groups in individual countries (e.g. COMA in the United Kingdom). Ideas of healthy eating by the public may not necessarily co-incide with established scientific fact. The demands for products described variously as 'organic', 'green' or 'natural' come from the consumer's own desire to establish eating habits consistent with his own conscience.

In producing animals to meet the needs of man, particularly for food, there has been a rapid development, in recent years, of materials which will modify the microbial population of the gut and supply nutrients from microbial biomass. Such efforts have been, on the one hand, to enhance animal performance and on the other, to do it without excessive use of materials such as antibiotics.

The science and practice of the industries involved are fast moving and much progress has been made in relatively few years. The authors believe that it is an appropriate time to review our knowledge in this field.

Dr Wesley Ewing
Dr Des Cole

CHAPTER 1

Introduction

The absorption of nutrients from the diet is the result of a series of complicated processes in the gastro-intestinal tract. In its simplest form it can be considered as the processes of hydrolysis and fermentation, with the end products of each being absorbed through the gut wall. The extent to which these processes take place, is influenced, for example, by the nature of the food, with the consequence that different foods have different digestibility values. Furthermore, the digestibility of the material will be influenced by the micro-flora which inhabits the digestive tract. The nutritionist, in many situations, seeks to modify these various processes in order to manipulate an animal's health and performance, to suit his needs.

As far as hydrolysis is concerned, the multitude of enzymes produced by the gut and present in the feed are specific to certain dietary components. In some situations a particular enzyme, for example amylase, is of very low activity at birth but will increase rapidly with age. In the new born pig, sucrase activity is non-existent and takes approximately 7 or 8 months to reach its maximum. In contrast, the lactase content of the gut is very high at birth, in order to meet the needs of the predominantly milk diet of the young animal, but its activity falls rapidly and, for example, in the pig has declined considerably by 5 weeks of age. (see *Figure 1.1*).

The gut micro-flora is inevitably linked to digestion and plays a particularly important role. In many species the micro-flora of the gastro-intestinal tract is quite substantial and most of these organisms are anaerobic with many being strictly obligate anaerobes. The colonisation of the mucosal epithelia by the micro-organisms is either by attachment to the epithelial cells or by their presence in the mucus layers at the base of the villi. By the nature of their position in the tract they are obviously closely associated with the digesta, and involved in digestion. In man and in the pig, micro-organisms are most numerous in the caecum and large intestine, where they play an important role in the fermentation processes. However, the site of fermentation varies depending on species.

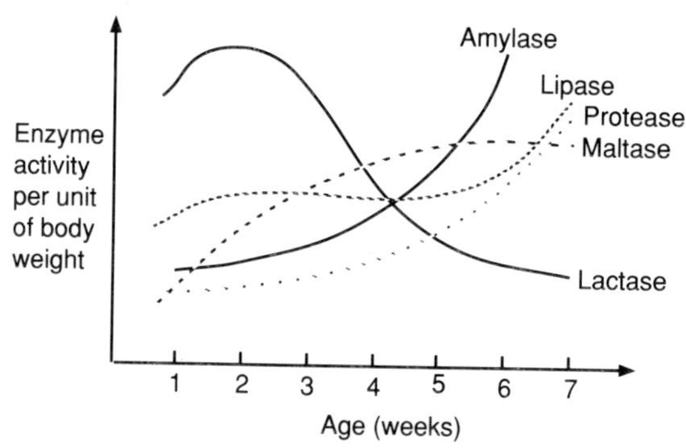

Figure 1.1 The development of digestive enzymes in a young pig. Micro-organisms present have to adapt to these changing conditions.

Man is constantly trying to change these various processes in the gut for his own needs. The extent to which he is successful will depend on many things, but it is important to recognise that while the site of breakdown is important, the site of absorption is more important. For example, while the horse has considerable ability to absorb the products of fermentation in the hind gut, the pig, although having considerable capacity for fermentation, is much less able to absorb from the hind gut.

It is not surprising that there have been many attempts to modify the health of the gut and therefore its efficiency in both man and his domestic animals. In this context, those developing animal products for the human market have played a particular role. The last decade has seen considerable challenges to the eating patterns of populations in many parts of the world. While cost of the diet is still a major consideration, safety and health within the context of perceived healthy eating have assumed major importance.

'Natural' and 'organic' foods play an increasing part in the advertising and marketing of food today. For example, some people are now prepared to pay a premium for guaranteed 'antibiotic free' meat produced in a 'natural' way. These changing attitudes and purchasing decisions will have a significant effect on the agricultural industry. This has already been seen in Great Britain during the 1989 Salmonella scare, and the 1990 BSE (Bovine Spongiform Encephalopathy) scare, when sales of eggs and beef respectively dropped drastically. Retailers today also have to meet public demand by supplying meat produced under the required conditions. Producers are now being asked to declare what they are feeding to animals, in order to regain the confidence of the public in eating meat.

With the intensification of the livestock industry there has been an increase in clinical and sub-clinical enteric disease. Thus, animals have become more vulnerable to harmful bacteria, such as *E. coli*, *Salmonella*, *Clostridium perfringens* and *Campylobacter sputorium*. As a response to the problems of intensification, there has been considerable reliance on antibiotics. Surveys (e.g. Smith and Halls, 1967) have shown that infective resistance is probably the most common form of drug resistance in *E. coli* inhabiting the alimentary tract of humans, calves, pigs and poultry. In 1970 Walton, in a survey of 400 pork and beef carcasses, found that the majority yielded drug resistant strains of *E. coli*.

A recent survey conducted by the Food Policy Research Unit at the University of Bradford examined the reasons (*Figure 1.2*) for consumers eating less meat. When offered eight different reasons which might have caused them to eat less meat, feed antibiotics, animal welfare issues, and the use of growth hormones were mentioned by approximately 10 per cent of consumers while the major reasons were cost and health (Woodward, 1988).

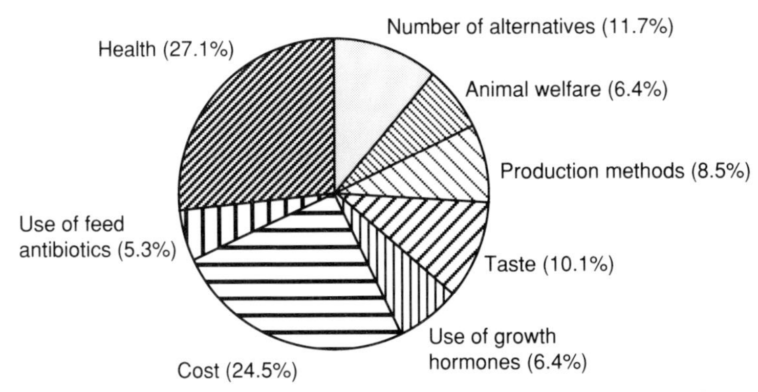

Figure 1.2 The reasons for eating less meat (% of total prompted responses) after Woodward (1988)

It has been shown that the administration of antibiotics to animals or humans at either therapeutic levels for a short time, or sub-therapeutic levels for a prolonged period, will increase the number of antibiotic resistant bacteria in the gastro-intestinal tract. However, whether, and to what extent, antibiotic resistant bacteria contribute detrimentally to human health has not been determined conclusively . While more than 2000 types of *Salmonella* have been identified, the strain associated with the egg scare in Britain was *Salmonella enteritidis* PT4. Equally important, however, is *Salmonella typhimurium* which is often found in animals' drinking water. *Salmonella*, like many other bacteria, can grow extremely quickly and, under optimum conditions, double in numbers every 15 minutes. Therefore, one strain of pathogen may be controlled while at the same time another develops.

In considering the activities within the gastro-intestinal tract, probiosis has received much recent attention. The word probiosis is originated from the Greek pro (for) and biosis (life), and has the opposite meaning of the term antibiotic. Instead of killing microbial cells a probiotic product is designed to promote the proliferation of the beneficial species of bacteria within the gut environment. In man, a typical example is the inclusion of yoghurt in the diet to modify gut flora. However, the greatest commercial development has been in farm animals and birds where preparations are fed to establish a desirable intestinal microbial balance and, consequently, to improve health and productivity. Such effects are achieved through the control of pathogenic organisms. Thus, the term probiotic could be applied to any material which can help to maintain the domination of pathogenic organisms by enhancing the activity of beneficial micro-organisms.

In this context all materials capable of a probiotic effect should be classified together. Some of the major ones can, for convenience, be considered as:

- Viable (live) bacteria.
- Non-viable (dead) bacteria.
- Live yeast.
- Dead yeast.
- Other fungi.
- Chemicals (e.g., lactic acid and other fermentation products).

Probiotics of a kind have been used throughout history, to modify or enhance biological processes and, for example, yoghurt has already been mentioned. The probiotic concept is therefore not new. In 1907 Metchnikoff stated that *horros autointoxicos* (the production of toxins by putrefying organisms) could be reduced by the continuous consumption of a fermented *Lactobacillus* culture. During the 1920's *Lactobacillus acidophilus* cultures continued to be used and it became known that it was one of the most common inhabitants of the intestinal tract of many animals. The concepts are still much the same today.

However, the probiotic definition has received much attention and been used to include those that are viable (live) cultures of bacteria, and those that are non-viable (dead) fermentation products (Pollmann, 1986). Fuller (1989) proposed that the definition should be a "live microbial feed supplement which beneficially affects the host animal by improving its intestinal microbial balance". This definition excludes fermentation products, e.g. lactic acid. Atherton and Robbins (1987) defined a probiotic as "any product which can help the normal flora to maintain their domination over pathogenic organisms" which fits well into the approach taken here. The word probiotic has been used by Lilley and Stillwell (1965) to describe substances produced by one protozoan which stimulated another, while Parker (1974) used it to describe organisms and substances which contribute to the intestinal microbial balance.

According to Chesson (1993) probiosis and probiotics should not be confused. He stated that probiosis is "the property of the normal adult flora to resist the overgrowth of component strains and the establishment of foreign strains", whereas "probiotics are intended to reinforce probiosis or re-establish probiosis where this has broken down because of environmental stress or as a result of extended treatment with antibiotics at therapeutic levels".

However, in contrast to Fuller (1989), we propose that a probiotic should be defined as any material (live, dead or chemical) which has a probiotic effect. There are also other materials of bacterial origin which do not have probiotic effects but supply beneficial nutrients to the animal and have their activity without involving probiosis. Conversely, there are materials of microbial and other natural origins which may influence performance of domestic livestock without having probiotic effects.

The increasing intensification of animal production systems, particularly within the pig and poultry sectors, has brought increased risks of both clinical and subclinical enteric disease. These can impair animal performance resulting in serious economic loss. In an attempt to control such problems. the use of in-feed antibiotics at both therapeutic and subtherapeutic levels has become widespread. However, growing concern over possible antibiotic residues in animal products has led to consumer pressure to reduce their use in feed. No matter whether this is entirely justified or not, it has led to increased interest in more natural ways of enhancing animal performance and helping the animal withstand disease.

Within this context, interest is now beginning to focus on one of the body's natural defence mechanisms which until now has received little serious attention. This is the population of non-pathogenic bacteria which are normally and naturally present within the gut of all domestic animals and birds as well as that of the human. Quite wrongly, most people have tended to think of bacteria as always being harmful. Only a very small proportion, however, fall into this category, with many being beneficial and essential to healthy living. There is a delicate balance of beneficial and pathogenic bacteria in the gastro-intestinal tract, and many symbiotic and competitive interactions occur between them.

Lactobacilli and other lactic acid bacteria are clearly important to man. This association has involved the manufacture of various human foods as well as including various beneficial interactions in different parts of the body. Renewed interest in the role of the intestinal micro-flora has focused on Lactobacilli, particularly *Lactobacillus acidophilus*, with lactic acid-producing bacteria assumed generally to be beneficial and non-pathogenic. They are very common organisms in nature, being found in milk, cheese and plant material, are involved in silage and yoghurt fermentation, and can be found in the saliva of man.

Certain species of *Lactobacilli*, when included in the diet, can have a beneficial effect for both humans and animals (Gilliland, 1979; Sandine, 1979) by helping to control the growth of undesirable micro-organisms in the intestinal tract. They have also been involved in the improvement of lactose utilisation in persons classified as lactose malabsorbers (Kim and Gilliland, 1983).

At the other side of this delicate balance are the pathogens, such as *Escherichia coli*, which cause disease and reach high numbers in the gastro-intestinal tract when the body's natural defence mechanisms (e.g. gastric acidity, other bacterial populations and antibody protection) are deficient or when either stress or infection reduces their protective effect. A variety of different factors may be considered to be "stresses" and result in a change in the balance of gut flora in favour of the pathogenic species. Live bacterial cultures have been claimed to stabilise this balance in favour of the beneficial bacteria, and therefore improve animal health and performance.

The concept of probiotics as applied to preventative medicine has been claimed to originate from Eli Metchnikoff in 1907. He postulated that the long life of some of the Balkan people was due to the regular consumption of fermented milk (yoghurt) containing *Lactobacillus bulgaricus*. Since then there have been many references to the application of probiotics in human health (e.g. Cheplin and Retger, 1922; Cheplin *et al.*, 1923; Kopeloff, 1926; Kopeloff and Beerman, 1925; Rettger, 1929; Retger *et al.*, 1935; Retger and Cheplin, 1921a, 1921b; Gotz *et al.*, 1979; Alm, 1983).

Metchnikoff's 'prolongation of life' was based on the ability of certain lactic acid bacteria to reduce the numbers of pathogenic gastro-intestinal micro-organisms. He stated that some microbes in the intestinal tract produced substances that were harmful to the host. Eating food containing the beneficial organisms, which he believed were contained in yoghurt, would improve the intestinal conditions. Through the constant addition of 'good' microbes in the diet, colonisation of the gastro-intestinal tract by disease-causing (pathogenic), 'bad' organisms, was prevented and thus health and life expectancy improved. This was the beginning of the probiotic concept of microbial inoculation based on the principle of competitive exclusion which has been confirmed by Crawford (1979). However, at the time his observations were treated with scepticism by some in the medical community. Since then he has been supported by workers such as Bohnhoff *et al.* (1954), who removed the natural micro-flora and by oral treatment of mice with a massive dose of streptomycin showed an increase in susceptibility to

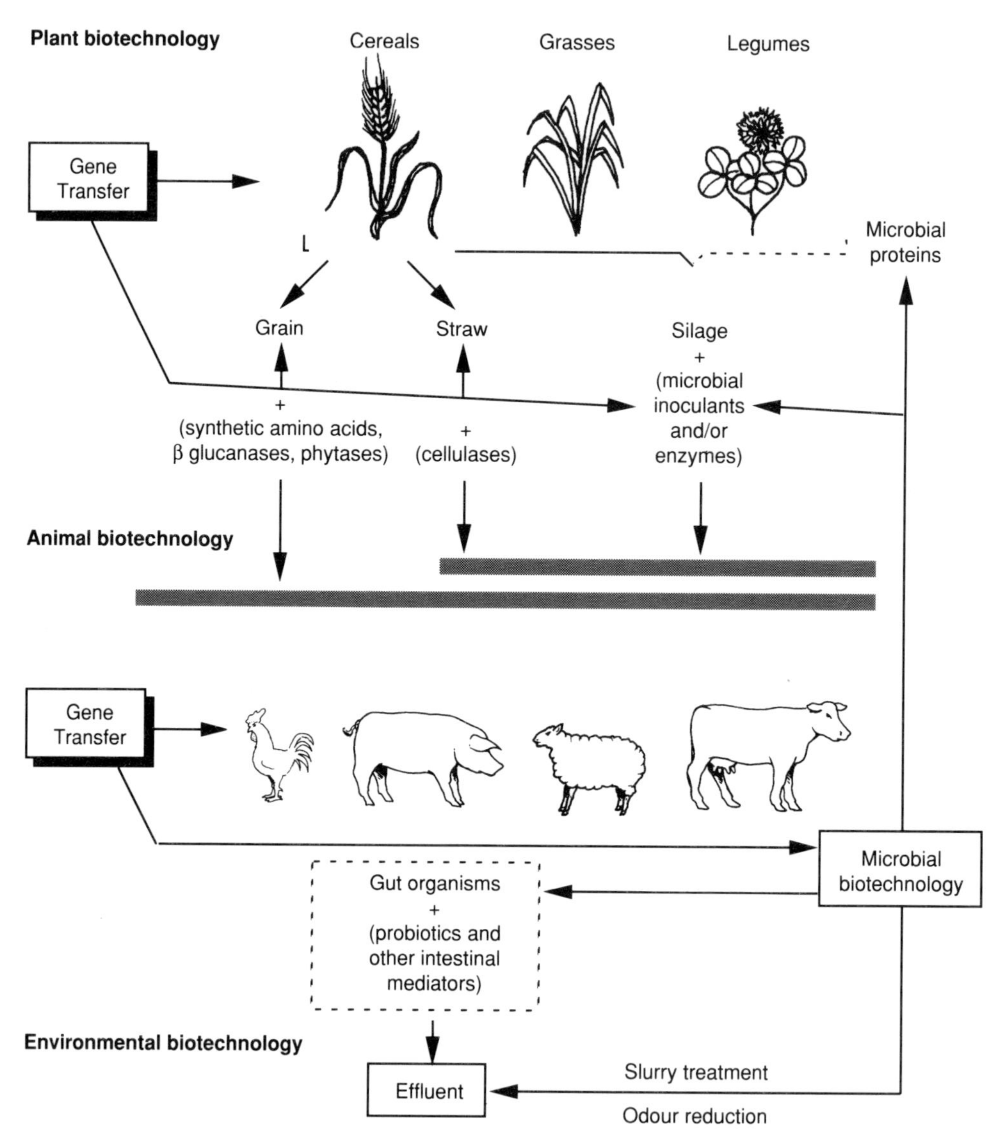

Figure 1.3 Biotechnology in the improvement of livestock food and feeding systems (after Robinson and McEvoy, 1993).

Salmonella typhimurium in the order of 100,000 fold. This was confirmed by later reports of Miller and Bohnhoff (1963) and Bohnhoff *et al.* (1964) who demonstrated the role of gut micro-flora in protection against a range of pathogenic enteric bacteria.

Probiosis has been claimed to provide a natural alternative to antibiotics or chemical growth promoters and it has been claimed that there is no problem of tissue residues and that the dependence on drugs can be reduced.

In the past, many researchers studying the effects of bacterial cultures have tended to accept *in vitro* results as confirmation of efficacy and others have been satisfied if the organism concerned simply attached to the gut lining. The animal feed industry is, however, concerned only with what can improve performance and health, while in human foods the emphasis is on health and well-being. Researchers have also concentrated on the use of live bacterial products for competitive exclusion. Today, an increasing number of reports show consistent effects in adult animals from feeding dead bacteria, or bacterial extracts.

Today's agriculture already relies on biotechnology in the improvement of livestock food and feeding systems (see *Figure 1.3*). Its involvement in a wide range of applications will undoubtedly grow as technologies develop. Some current applications are listed below.

- Phytase which improves the digestibility and availability of phosphorus in poultry and pig diets (Beers and Jongbloed, 1992; Ketaren *et al.*, 1993).
- Enzyme preparations containing β–glucanase and arabinoxylanase activities for addition to barley and rye based diets for broilers, to improve performance (Patterson *et al.*, 1990).
- Mixed enzyme preparations in pig diets in enhancing growth rate and preventing enteric disease (Close, 1992).
- Modified *Lactobacillus* preparations in controlling silage fermentation (Sharpe *et al.*, 1993).
- Gene manipulation to allow the enhancement of UDP levels in forage plants (Rogers, 1990).
- Gene manipulation to allow the monogastric animal to improve digestion.
- Oligosaccharides to control bacterial overgrowth (Chesson, 1993).
- Bacterial additives as probiotics (Fuller, 1989).
- Fungal probiotics to enhance fibre breakdown and protein flow to the abomasum (Wallace and Newbold, 1992).
- Manipulated bacteria to degrade plant toxins (Jones and Megarrity, 1986).

CHAPTER 2

The gastro-intestinal tract

Introduction

The complexities of the gastro-intestinal tract differ from species to species, and site to site within the gut and depend on the health of the animal. The key to digestion is the absorption of nutrients through the gut wall in a usable form. In rendering food suitable for absorption, the mechanisms can be thought of in the simplest way as hydrolysis and fermentation.

Man, pigs and poultry are non-ruminant, simple stomached animals (see *Figures* 2.1 and 2.2), and do not rely on a symbiotic relationship with the microbial flora within their gastro-intestinal tract to survive to the same degree as animals possessing a rumen. Their gastro-intestinal bacterial populations are further down the tract than ruminants, so they do not ferment food to break down indigestible material to such a large extent. A smaller micro-flora compared with ruminants means that they have less digestive ability and consequently the nutritional demands of their diet is increased. They do not cope as well with poor quality feedstuffs containing large quantities of cellulose and lignin and structural carbohydrates, and have a considerable reliance on the diet for amino-acids and vitamins.

Ruminant animals, however, have considerable microbial fermentation occurring within the gastro-intestinal tract (*Figure* 2.3). This fermentation takes place in the rumen before digestion in the small and large intestines. These animals have the ability to synthesise amino-acids and most vitamins through microbial action on nitrogen and carbon sources. Also, unlike pigs and poultry, they have digestive enzymes which allow the breakdown of structural carbohydrates (*Figures* 2.5 and 2.6). For example, hydrolysis of beta-linked polysaccharides occurs, making the digestion of roughage possible. Horses, although herbivores, do not have a rumen but have an enlarged caecum (*Figure* 2.4).

The heavy microbial load in the rumen, just like the host, requires nutrients to survive, and there is a cost associated with this increased digestive capacity. In addition,

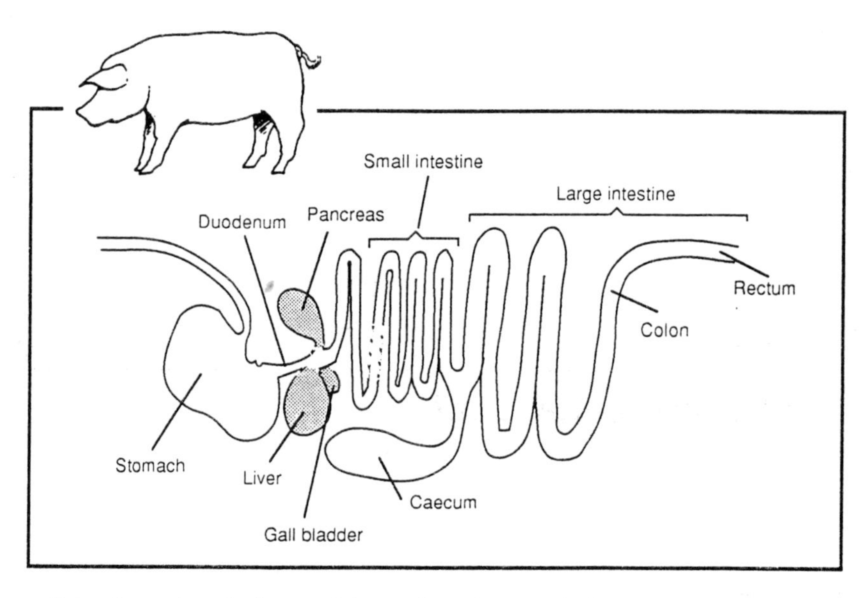

Figure 2.1 Gastro-intestinal tract of the pig (simple stomached).

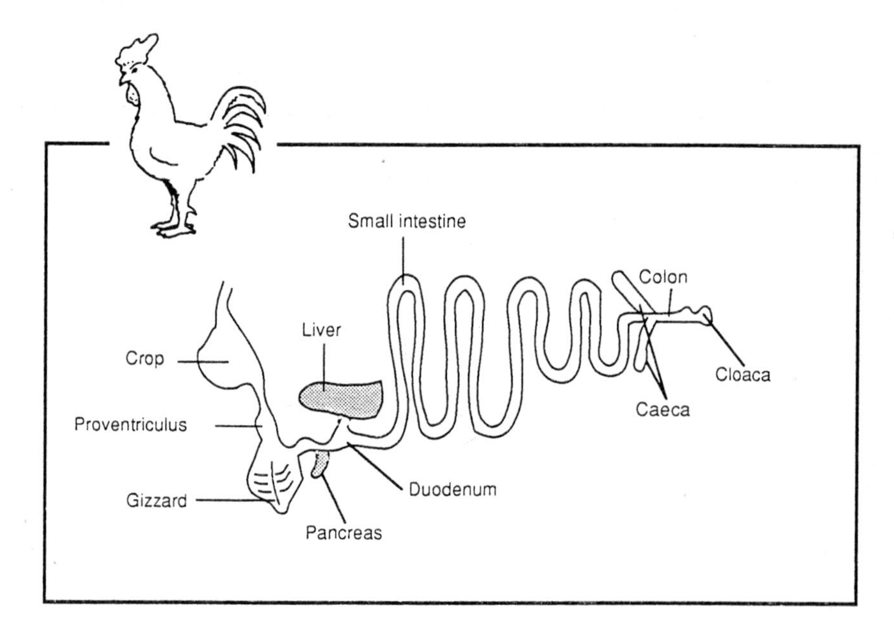

Figure 2.2 Gastro-intestinal tract of poultry (simple stomached).

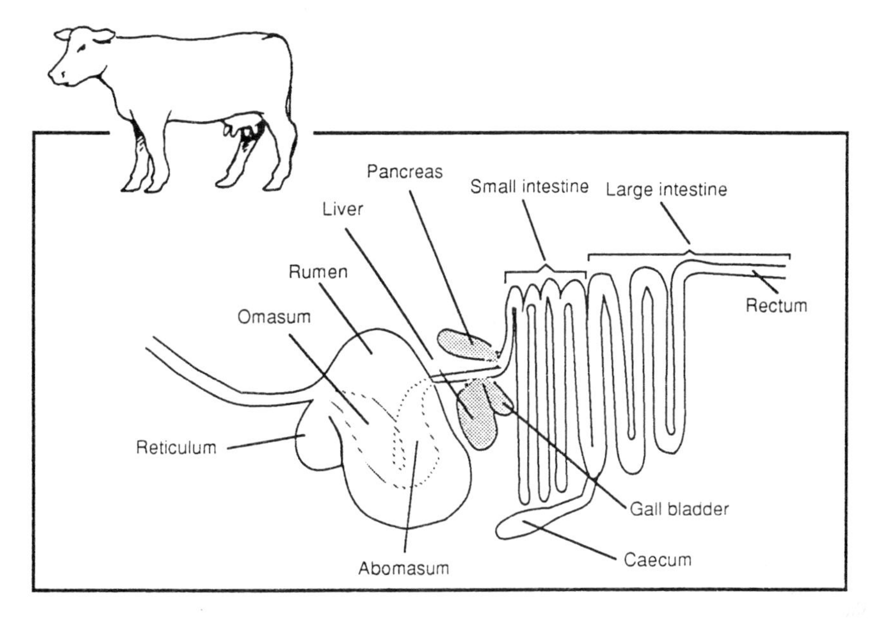

Figure 2.3 Gastro-intestinal tract of the cow (ruminant).

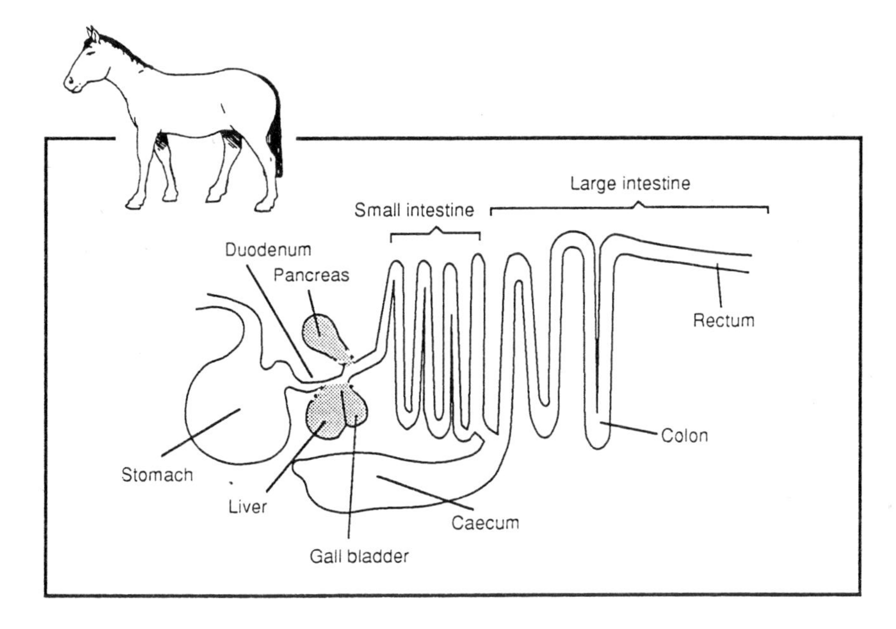

Figure 2.4 Gastro-intestinal tract of the horse (simple stomached non-ruminant herbivore).

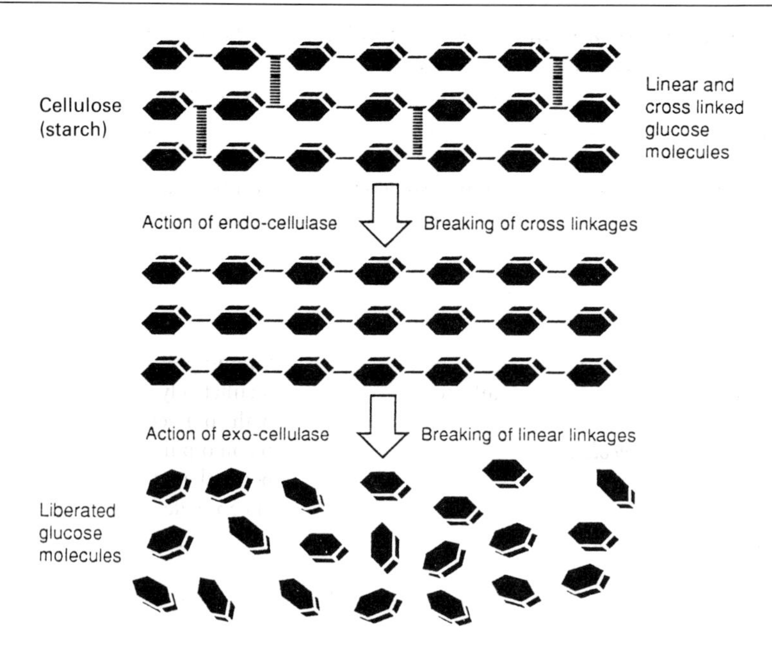

Figure 2.5 Illustration of action of cellulases on starch with the final liberaton of glucose molecules.

when high quality feeds are given, this microbial maintenance demand means the animal converts high quality feedstuffs into body tissue, or milk, less efficiently than would a non-ruminant. The microbial protein may also be of lower quality than the original high quality feed.

A wide array of bacteria have been isolated from healthy animals, most of which are natural to the environment in which they live. Donaldson, in 1964, detailed the bacteria of the intestines and described them and their distribution in the environment (Table 2.1).

Organs and digestion

Different organs assume various degrees of importance in different species (*Figures 2.1–2.4*) and have different enzymic and digestive activity (see Table 2.2). Consequently a short description of the important organs is necessary.

Monogastric animals

The mouth

The action here is largely mechanical with the particle size being reduced by mastication. Saliva, secreted by the salivary glands, has a lubricating effect on the food. In mammals

Table 2.1 *Bacteria isolated from the intestinal tracts of healthy animals.*

Bacterium	*Distribution*	*Description*
Bacteroides: *Bacteroides funduliformis*, *B. fragilis*, *B. putidus*, *B. pneumosintes*, *B. serpens*, *N. nigrescens*	Intestinal tract; mouth; respiratory tract.	Gram-negative, strictly anaerobic, nonsporulating, frequently encapsulated slender rods, with many bizarre pleomorphic forms.
Enterobacteria (coli-aerogenes) *Escherichia coli*, *E. freundii*, *Alkalescens-dispar paracolon*, *Aerobacteria aerogenes*, *A. cloacae*, *Klebsiella rhinoscleromatis*, *K. oxytoca*	Lower intestine; plants; soil.	Gram-negative, aerobic or facultatively anaerobic small rods, with rounded ends, several species pleomorphic; motile species have peritrichate flagella; frequently encapsulated, but nonsporulating.
Enterobacteria (proteus-providence): *Proteus vulgaris*, *P. morganii*, *P. mirabilis*, *P. rettgeri*	Lower intestine; soil; sewage; plants.	Gram-negative, aerobic or facultatively anaerobic rods varying from 1 to 30 microns in length; no capsules or spores formed.
Enterococci: *Streptococcus faecalis*; *Var. zymogenes*, *Var. liquefaciens*, *S. faecium*, *S. durans*, *S. bovis*, *S. lactis*	Intestinal tract; mouth; sewage; vegetation.	Gram-positive, aerobic or facultatively anaerobic small cocci in pairs or short chains.
Lactobacilli: *Lactobacillus acidophilus*, *L. bifidus*, *L. brevis*, *L. exilis*, *L. casei*	Milk; intestinal tract (infants); mouth; stomach; vagina.	Gram-positive, micro-aerophilic or anaerobic long slender rods, which are often pleomorphic & occur singly or in chains.
Pseudomonads: *Pseudomonads aeruginosa*, *Ps. fluorescens*, *Ps. ovalis*, *Alcaligenes faecalis*	Intestinal tract; sewage, water.	Gram-negative, aerobic slender rods occurring singly in short chains or in small bundles; polar flagellae.
Clostridia: *Clostridium perfringens*, *C. tetani*, *C. botulinum*, *C. fibermentans*, *C. sporogenes*	Soil; dust; intestinal tract.	Gram-positive, anaerobic, spore-forming, large, broad rods; marked pleomorphism; frequently encapsulated; peritrichate flagella in motile species.
Bacillus subtilis		

(Donaldson, 1964)

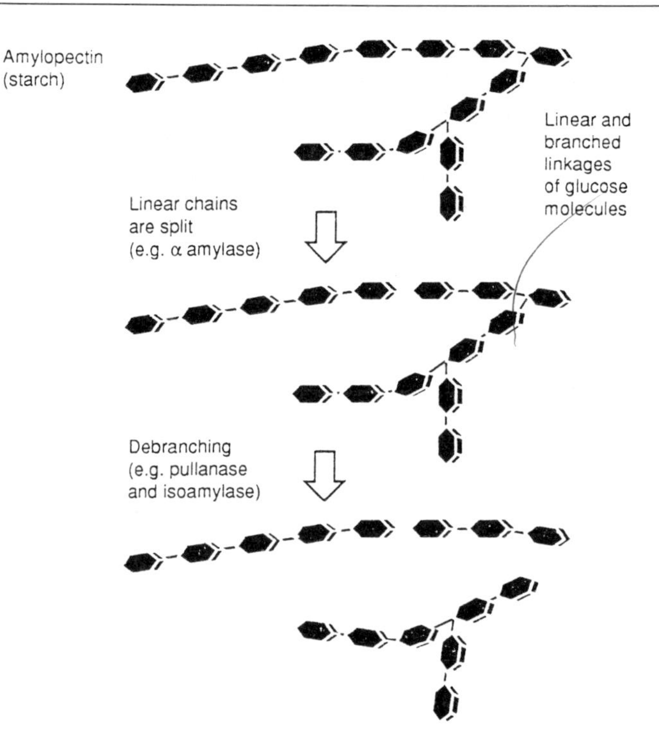

Figure 2.6 The major starch degrading enzymes include α and β amylases, gluco amylases, pullanases and isoamylases. Starch exists in two forms, amylose (linear) and amylopectin (branched). Amylase hydrolyses α 1, 4 bonds, while examples of debranching enzymes (which break the α 1, 6 linkages) are pullanase and isomylase.

it contains mucin, water and in some herbivores and man, the amylase ptyalin, which catalyses the breakdown of starch to maltose. The extent of amylase activity in the mouth is variable, being high in man and lacking in cats, dogs and horses. Little starch digestion is thought to occur in the mouth due to an unfavourable pH.

Stomach

The stomach is an organ for digestion and storage. In the adult pig, it has a capacity of about 8 litres. While it is a single compartment, there are four main zones in the stomach of the pig, which is representative of other single stomached mammals, and each has a different structure, environment and therefore microbial loading and type. The organ has a very low pH, necessary for proteolysis, and consequently microbial activity is very much reduced.

The four zones are oesophageal, cardiac, gastric or peptic and pylorus (*Figure* 2.7). The hormone pylorin, produced in the pylorus, stimulates the production of gastric juice which contains, in addition to water, hydrochloric acid, pepsinogens, inorganic salts, mucus and the factor involved in the absorption of vitamin B_{12}. The pH of the stomach of the pig is about 2. The pepsins involved have two pH optima of 2.0 and 3.5. Protease enzymes within the digestive juices break the polypeptides into shorter

Table 2.2 *Source and action of digestive enzymes.*

Source	*Enzyme/digestive secretion*	*Action*
Salivary glands (mouth)	Amylase	Glycogen, dextrin and starch broken to branched oligosaccharides and maltose
Gastric glands (stomach)	Pepsinogen	Causes hydrolysis of peptide bonds in polypeptides
	Histamine	Causes proteins to swell. Antibacterial effect
Pancreatic glands (pancreas)	Tryposinogen	Breaks protein to polypeptides
	Chymotrypsinogen	Breaks protein to polypeptides
	Procarboxypeptidase	Breaks polypeptides to small peptides and amino acids
	Carboxypeptidase	Breaks polypeptides to small peptides and amino acids
	Elastase	Hydrolyses fibrous proteins
	Collagenase	Hydrolyses collagen
	Lipase	Attacks fats/glycerol
	Phospholipase A	Degrades phospholin by removing a hydrocarbon
	Cholesterol esterase	Attacks cholesterol
Glands in mucosal microvilli (small intestine)	Aminopeptidases	Breaks polypeptides to small peptides and amino acids
	Dipeptidases	Breaks dipeptides to amino acids
	Phosphatase	Attacks organic phosphates
	Monoglyceride lipase	Attacks monoglyceride fatty acids and glycerol
	Lecithinase	Lecithin to fatty acids, phosphoric acid
	Sucrase	Sucrose to glucose, fructose
	Maltase	Maltose to glucose
	Lactase	Lactose to glucose, galactose
Gall bladder (liver)	Bile	Emulsifies fat Stabilises emulsions Neutralises acid chyme Accelerates action of pancreatic lipase pigment and cholesterol excretion

amino acid chains (see *Figures 2.8–2.10*). Eventually the protein is broken down to single peptides or amino acids.

The oesophageal region which joins directly from the oesophagus has no glandular secretions, unlike other parts of the stomach. Its surface is very similar to that of the crop of birds, with the associated micro-flora comprising mainly *Lactobacilli*. *Lactobacilli* can

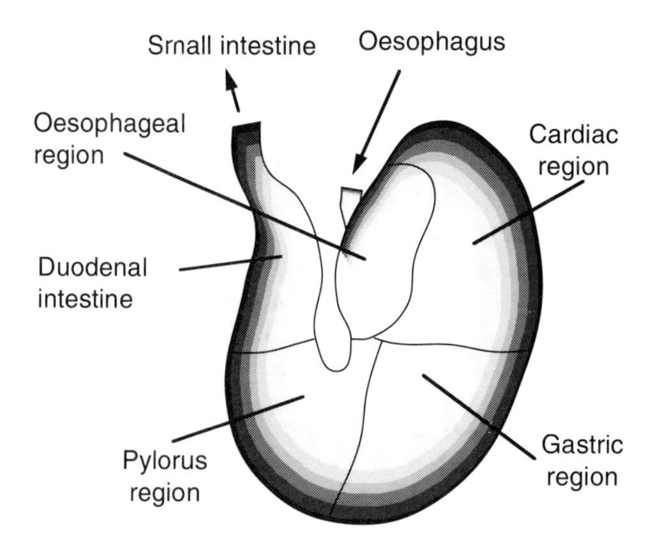

Figure 2.7 The regions of a pig's stomach.

be found here at all ages in the animal's life (Tannock and Smith, 1970). The type and number of bacteria, specifically *Lactobacilli* and *Streptococci*, have been shown to depend on their ability to adhere to and colonise the surface. Common bacterial types include *L. fermentum*, *L. salvaricus*, and *S. salivarus* in younger pigs, and *L. acidophilus* and *S.bovis* in older animals (Fuller *et al.*, 1978). A low level of *L. acidophilus* in older animals has been associated with enteritis (Redmond and Moore, 1965). As the pig gets older, pepsin and HCl are produced in larger quantities. Milk-fed

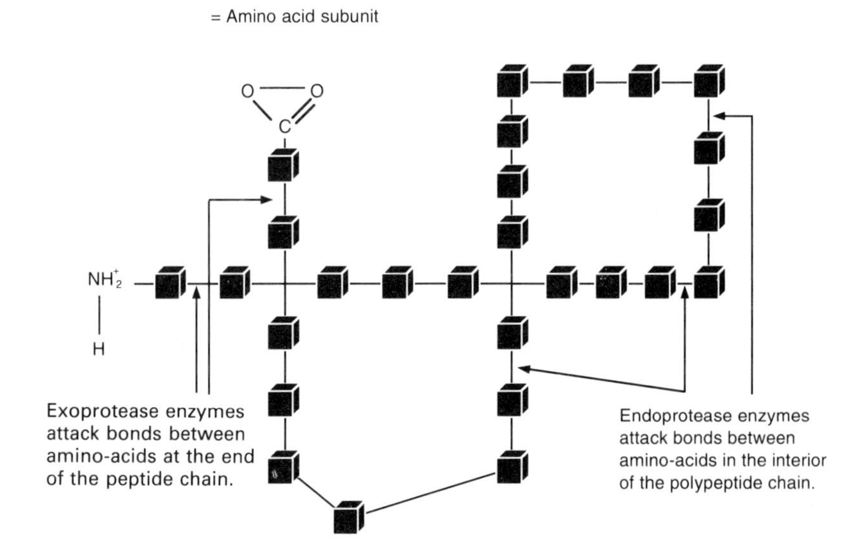

Figure 2.8 The break-down of protein in the food by the action of protease enzymes.

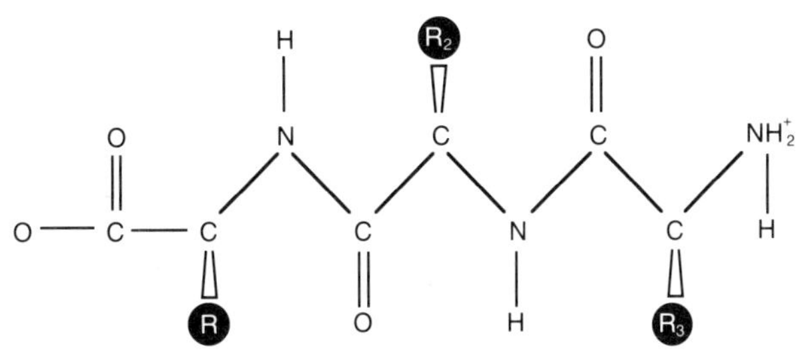

Figure 2.9 Illustration of the structure of a tri-peptide chain of amino acids. R = Functional group. N.B. A polypeptide is a chain of 10 or more amino acids.

piglets have a stomach pH in excess of 3.4, while at 2–3 weeks of age there is a much lower pH. This will reduce the level of *L. acidophilus*.

The stomach of the animal is affected by the conditions in which it lives. When a high microbial load is present, the level of luminal fermentation increases while gastric digestion is delayed in pigs. The pH of the stomach is inversely related to the microbial load in the animal's environment. The resultant microbial fermentation leads to the production of organic acids which act as a buffer, demanding further HCl to be secreted to lower the pH for pepsinogen activity. When dietary protein is high, its buffering capacity can prevent the pig's stomach achieving a low pH.

Digestion at the upper portion of the stomach (oesophogeal area) is unaffected by gastric juice produced lower in the organ. This allows favourable pH, moisture, temperature and substrates for microbial development and fermentation. The extent of fermentation also depends on the composition of the feed (i.e. lactose, sucrose, etc., produce higher levels of fermentation than starch). Other factors include particle size in the feed and the extent of starch gelatinisation after pelleting.

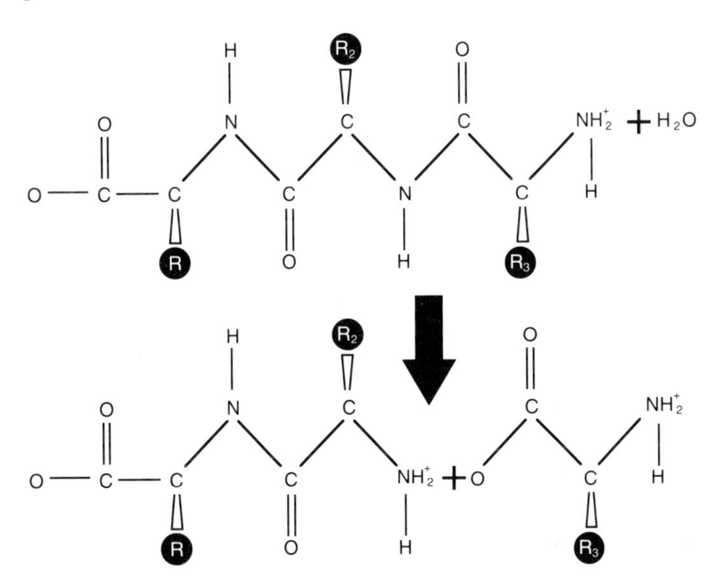

Figure 2.10 Illustration showing partial hydrolysis of a tri-peptide to a di-peptide plus a single peptide.

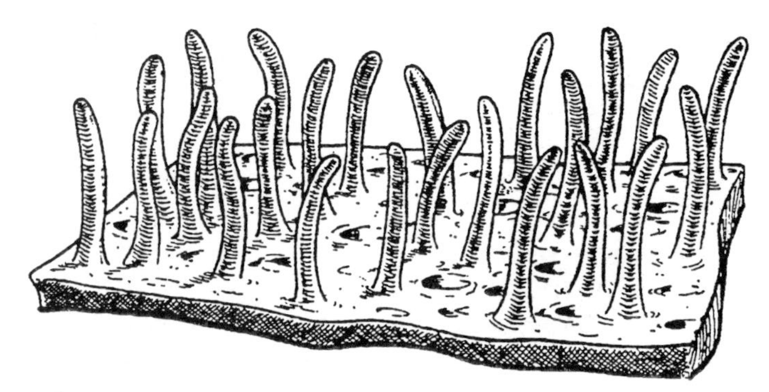

Figure 2.11 Illustration of villi on intestinal surface.

As the food moves down to the gastric region, the pH is lower, microbial activity decreases, and pepsin activity increases.

The small intestine

This long, convoluted organ starts at the pylorus and ends at the ileo-caecal junction. It can be up to 20 metres in length in the pig and account for one third of the volume of the gastro-intestinal tract. It is made up of three areas: the duodenum, jejunum and ileum; but the jejunum accounts for about 85% of its length. Food is partly digested by the time it reaches the small intestine. The digesta becomes mixed with the secretions of the duodenum, liver and pancreas.

The small intestine is an important site of digestion, largely by the host enzymes but, to a certain extent, by fermentation. It is populated with a micro-flora which increases in population as the large intestine is approached. It is also a major site of absorption of the end products of the digestive processes. Absorption is made easier by the great increase in surface area which is provided by the villi in the small intestine (*Figure 2.11*). *Plates 2.1* and 2.2 show the villi of pigs pre- and post-weaning. The villi of pigs post weaning can be seen to be slightly stressed and reduced in length.

The large intestine

The large intestine comprises the colon and caecum. Most of the nutrients capable of being hydrolysed will have been absorbed by the time the digesta reaches the large intestine. The largely cellulose and various hemicellulose materials are subjected to a certain degree of breakdown by the large microbial population, which includes *Lactobacilli*, *Streptococci*, coliforms, *Bacteroides*, *Clostridia* and yeast. The end products of fermentation include acetic, propionic and butyric acids, together with indole, skatole, phenol, hydrogen sulphide, amines and ammonia.

The extent to which the end products of digestion are absorbed is open to question. While there is a certain amount of absorption of volatile fatty acids, the primary function of the caecum and colon is to maintain the water and electrolyte balance. Table 2.3 illustrates the relative volumes and lengths of the main regions of the pig's digestive tract.

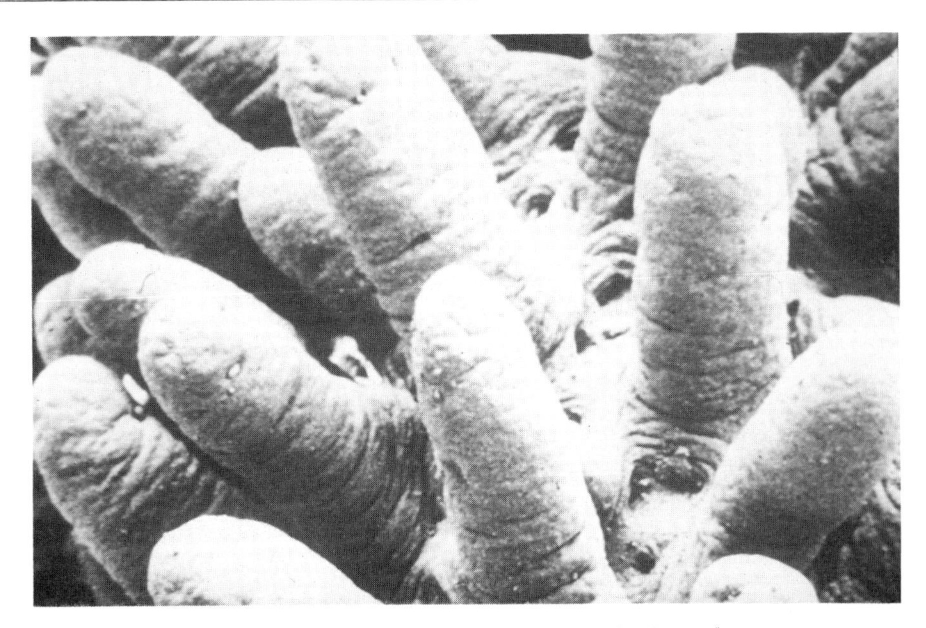

Plate 2.1 Healthy villi in pigs pre-weaning. [*By courtesy of John Bourne*]

Plate 2.2 Shortened stressed villi in pigs post-weaning. [*By courtesy of John Bourne*]

Table 2.3 *Typical volumes and length of main regions of the pig's digestive tract.*

	Volume (ℓ)	*Length (m)*
Stomach	9	–
Small intestine	11.5	18.3
Large intestine	9	4.9

The large intestine allows animals to make full use of their food by the salvaging of energy from dietary carbohydrates (eg. resistant starches, non starch polysaccharides (dietary fibre), sugars and oligosaccharides) which have been neither digested nor absorbed in the small intestine (Gibbons *et al.*, 1993). This process relies on anaerobic bacteria to breakdown the carbohydrates to short chain fatty acids by fermentation. Fermentation can also help utilise dietary proteins and host derived substances, such as pancreatic enzymes, mucus and sloughed off epithelial cells (Cumming and Macfarlane, 1991). The entire process relies on the natural production of hydrolytic enzymes by anaerobic bacteria.

pH

The pH of the gut varies dramatically from the mouth and stomach to the colon and caecum. The stomach is kept highly acidic due to the release of hydrochloric acid, protease and mucus bile and can be as low as pH 1.5. However, the pH is usually around 2.5–4.5 for adult animals (specifically for pigs) and 4.5–7 for young unweaned pigs. The stomach is highly oxidised, thus further limiting microbial growth.

The caecum and colon are more neutral at pH 6–8 and are consequently the site of a lot of microbial activity. Table 2.4 indicates the approximate pH of the different regions of a pig's digestive tract. A higher pH will give gut pathogens such as *E. coli* the opportunity to multiply at a quicker rate. Table 2.5 shows the typical pH range for the growth of common micro-organisms. Some micro-organisms are acid tolerant (acidophilic), while others are alkaline tolerant (alkalinophilic).

Table 2.4 *pH of contents of various regions of pig's digestive tract.*

Region	*pH*
Stomach	1.5–6.0
Duodenum	6.0–8.5
Caecum	6.0–9.0
Rest of large intestine	8.0–9.0

Table 2.5 *The pH ranges for growth of common micro-organisms.*

Organism	*Minimum*	*Optimum*	*Maximum*
Examples of types:			
Bacteria	3.0–4.0	6.0–7.5	9.0–10.0
Yeasts	2.0–3.0	4.5–5.5	7.0–8.0
Moulds	1.0–2.0	4.5–5.5	7.0–8.0

Poultry

The digestive system of poultry differs considerably from other monogastric animals, (*Figure* 2.2). For example, it starts with a beak, and does not have teeth for mastication of food.

A further major difference is that the chicken has a crop (diverticulum) which is a pear-shaped reservoir in the oesophagus for holding food. There is some microbial activity in this organ with the production of lactic and acetic acids. The crop is linked by the proventriculus to the gizzard which is similar in function to the pyloric part of the pig stomach.

Starch enters the crop, which is a storage organ, and is hydrolysed by microbial activity, but sucrose does not undergo any microbial action. The production of volatile fatty acids in the crop indicates that some of this bacterial activity is from anaerobes. However, their level is not always high. *Lactobacilli* are the major bacterial inhabitants and are involved in the correct maintenance of microbial balance. This balance is affected by stress (see Chapter 4). Antibiotics can affect both the *Lactobacillus* level and the total microbial biomass, with a reduction in the total level being related to improvement in performance.

The small intestine of poultry is much shorter than that of pigs but still has a significant micro-flora. Poultry have large caeca and a small colon. The caeca assume the role of mixing the digesta together with fermentation and absorption. *Figure* 2.12 illustrates the major microbes in the main areas of the digestive tract of pigs and poultry.

Ruminants

The major characteristic of this group of animals is the presence in the adult of a large organ of considerable fermentative capacity. It is divided into four compartments, the rumen, the reticulum, the omasum and the abomasum. Table 2.6 details the typical maximum volume and length of the regions of the cow's digestive tract.

The juvenile animal, while suckling, behaves like a monogastric animal. There is little development of the rumen and reticulum whilst suckling, with milk passing down the oesophageal groove into the omasum and abomasum. The groove is formed as a reflex action by the oesophagus closing when milk is drunk from a teat.

As the young ruminant eats solid food there is considerable development of the rumen and reticulum. Eventually, they occupy more than 80% of the capacity of the total

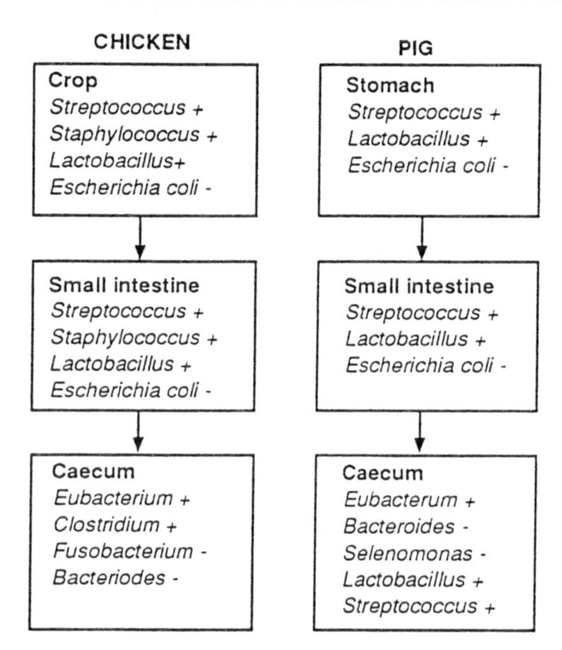

Figure 2.12 Micro-organisms in the main areas of microbial activity of the digestive tract. Gram-positive (+) and Gram-negative (–)

stomach, with the oesophageal groove no longer in effect. Saliva plays an important part in the dilution of food. In the rumen, food is broken down physically and chemically. The rhythmic contractions of the rumen walls mix the food, which is regurgitated to be chewed and swallowed again.

The chemical breakdown which occurs in the reticulo-rumen results from the enzymes produced by anaerobic bacteria, protozoa and fungi (Table 2.7). The resulting fermentation produces volatile fatty acids, methane and carbon dioxide with the rumen usually functioning at a pH of about 5.5–6.5. A summary of the action of the rumen on food is given in *Figure 2.13*. Many species of bacteria have been identified in the rumen (Table 2.8) and are found at about 10^9–10^{11}/ml of rumen

Table 2.6 *Typical volumes and length of main regions of the cow's digestive tract.*

	Volume (ℓ)	*Length (m)*
Rumen	182	–
Reticulum	23	–
Omasum	68	–
Abomasum	32	–
Small intestine	91	45
Caecum	11	1
Large intestine	32	10

Table 2.7 *Products of rumen microbial fermentation.*

Type of Organism	*Product*
Cellulose fermenting	Acetic, succinic and lactic acid
Starch and sugar fermenting	Lactic, butyric, formic, succinic, propionic and acetic acids
Lactic acid and succinic fermenting microbes	Propionic and acetic acid
Vitamin producing microbes	Vitamin B and others

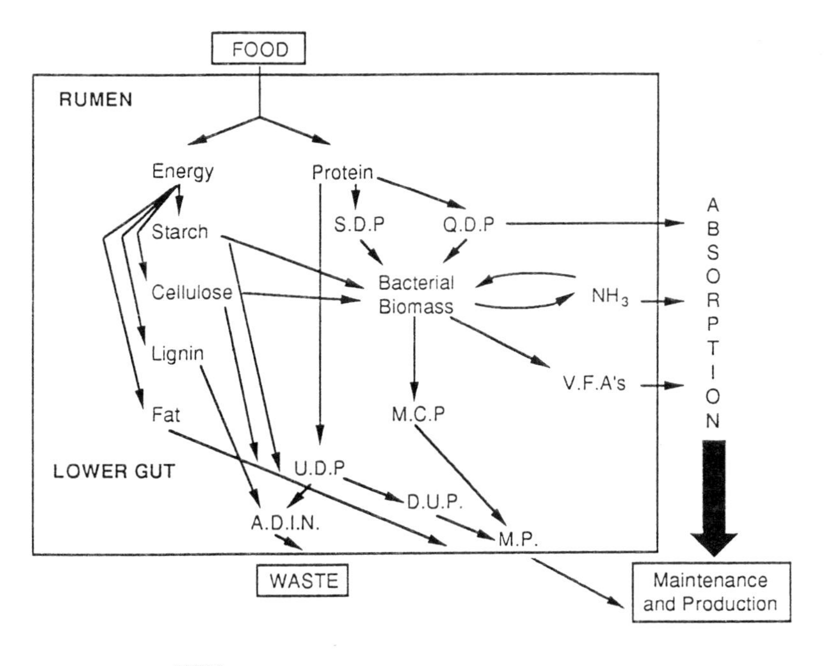

KEY

S.D.P = Slowly degraded protein
Q.D.P. = Quickly degraded protein
NH3 = Ammonia
V.F.A.'s = Volatile fatty acids
M.C.P. = Microbial crude protein
U.D.P. = Undegraded protein
D.U.P = Digestible undegraded protein
A.D.I.N. = Acid detergent insoluble nitrogen
M.P. = Metabolisable protein

Figure 2.13 The breakdown of food in the rumen.

Table 2.8 *Common bacteria of the rumen.*

Type of Organism	*Name*
Cellulose-fermenting	*Bacteroides succinogenes* *Ruminococcus flavefaciens* *Butyrivibrio fibrisogenes*
Hemicellulose fermenting	*Bacteriodes ruminicola* *Ruminococcus flavefaciens* *Butyrivibrio fibrisogenes* *Ruminococcus albus*
Starch and sugar fermenting	*Streptococcus bovis* *Butyrivibrio fibrisolovens* *Succinivibrio dextrinosolens* *Bacteroides ruminicola* *Selenomonas ruminantium*
Lactic acid and succinic acid fermenting	*Veillonella gazogenes*
Vitamin synthesising	*Flavobacterium vitarumen* *Clostridium butyricum*

content. Protozoa (e.g. *Isotrichia, Dasytricha, Entodinium, Diplodinium, Epidinium* and *Ophryosolex*) are present in much smaller numbers.

The epithelial tissue of the rumen is a very special environment, because large amounts of urea diffuse through the rumen wall and must be transformed to ammonia to avoid toxic effects. The taxonomically distinct population of approximately 23 bacterial species that colonise the rumen wall includes several species that produce large amounts of urease, with this enzyme essential for normal physiological function. The mature rumen is also host to a large population of yeast (fungi). Yeast grow by budding and dividing which involves the formation of a bud on the mother cell which grows to the size of the mother and then the cells separate. After division, the cells are distinguishable, as the mother cell retains a bud scar for each daughter cell produced. A population of yeast cells which is growing has a changing cell age distribution. Under some growth conditions the buds do not separate from the mother cell and long, branched chains of cells (pseudomycelia) will form. Under optimum conditions yeast may divide in as little as 45 minutes, but generally it takes at least twice this time.

Sites of microbial fermentation

The form of the gastro-intestinal tract will determine the nature of its digestion (see *Figures 2.1–2.4*). While ruminants have considerable capacity for fermentation in the rumen, some simple stomached animals have well developed hindguts.

Table 2.9 *Classification of some animals based on gastro-intestinal anatomy.*

Class	*Species*	*Dietary habit*
Pregastric fermenters		
Ruminants	Cattle, sheep, deer	Grazing herbivores
	Antelope, camel	Selective herbivores
Non-ruminants	Colobine monkey	Selective herbivore
	Hamster, vole	Selective herbivore
	Kangaroo, hippopotamus.	Grazing and selective herbivores
	Hoatzin	Folivore
Hindgut fermenters		
Caecal	Capybara	Grazer
	Rabbit	Selective herbivore
	Rat	Omnivore
Colonic digesters		
Sacculated	Horse, zebra	Grazer
	New world monkey	Folivore
	Pig, man	Omnivores
Unsacculated	Dog	Carnivore
	Cat	Carnivore

(Van Soest, 1991)

The main sites of microbial fermentation differ from species to species. Animals have been classified on this basis (Table 2.9) and can be considered in the simplest form to be pre-gastric or hindgut fermenters, depending on the major site of fermentation.

The herbivores are largely the pre-gastric fermenters. Ruminants, such as cattle, sheep and goats, are particularly well developed for microbial fermentation, with the multi compartment stomach retaining food for a long time. There are non-ruminant pre-gastric fermenters which also have large, complex stomachs, but they do not ruminate or regurgitate their food for secondary chewing.

Hindgut fermenters rely primarily on the caecum and colon for microbial fermentation. Some herbivores (e.g. the rabbit) have a very well developed caecum which accounts for more than 40% of the digestive tract. However, the rabbit is unusual in that coprophagy is an important aspect of its nutrition. Others, such as the horse and pig, rely to a greater extent on the colon. Animals vary considerably in their ability to absorb nutrients digested at this level of the digestive tract. The horse has considerable ability, while some species have little.

Microbial fermentation results in the liberation of different fermentation products, depending on the organism (see Table 2.10). For example *Lactobacillus* produces lactic acid from pyruvate, while *Saccharomyces* liberates carbon dioxide and ethanol.

The relative fermentative capacities of different species vary considerably, as do the various parts of the digestive tract in their capacity for fermentation. (Table 2.11).

Table 2.10 *The variation in fermentation products produced by the metabolism of pyruvate by different micro-organisms.*

Organism	*Fermentation product*
Lactobacillus	Lactic acid
Streptococcus	Lactic acid
Clostridium	Acetone, butyric acid, butanol, isopropanol
Saccharomyces (Yeast)	Carbon dioxide, ethanol

Furthermore, the strict division into pre-gastric and hindgut fermenters may be too simplistic. For example, it has been shown in the pig, which is classified as a hindgut fermenter, that considerable fermentation may have taken place before the digesta reaches the large intestine. In one example, 9.3% of the crude fibre and 17.9% of the non-starch polysaccharides had disappeared before the ileum, with a further 30.0% and 49.2% being absorbed from the large intestine (Table 2.12).

Taking pigs as a further example, some fermentation of carbohydrates appears to begin in the first few days of life with activity throughout the gut even at this early age. The potential substrates for this fermentation are any of the carbohydrates consumed by pigs and it is likely that both microbial and host enzymic activity continue in parallel in the whole of the stomach and small intestine. By the end of the small intestine, the

Table 2.11 *Fermentative capacity expressed as percentage of the total digestive tract.*

Species	*Reticulo-rumen*	*Caecum*	*Colon and rectum*	*Total fermentative capacity*
Cattle	64	5	5–8	75
Sheep	71	8	4	83
Horse	–	15	54	69
Pig	–	15	54	69
Capybara	–	71	9	80
Guineapig	–	46	20	66
Rabbit	–	43	8	51
Man	–	32	29	61
Man	–	–	17	17
Cat	–	–	16	16
Dog	–	1	13	14

(Parra, 1978)

Table 2.12 *Examples of digestion and absorption in different parts of the digestive tract of the pig.*

	Starch (%)	*Crude fibre (%)*	*Non-starch polysaccharides (%)*	*Fatty acids (%)*
Diet	100	100	100	100
Disappearance				
by ileum	97.2	9.3	17.9	72.6
in large intestine	2.5	30.0	49.2	8.8

(Dierick, 1991)

Table 2.13 Bacillus *organisms produce a wide variety of enzymes as listed below.*

Organim	*Enzyme*
B. amylolique faciens *B. licheniformis*	α-amylase
B. coagulans	Glucose isomerase
B. amylolique faciens *B. circulans* *B. subtilis*	β–glucanase
B. stearothermophilus	Exo-amylase
B. licheniforms	Alkaline protease
B. amylolique faciens	Neutral protease

amounts of endogenous amylase present are very low and microbial enzyme activities, including proteases, high. However, the concentrations and activities of enzymes from microbial and endogenous sources show that endogenous activity predominates in the stomach and much of the small intestine of the pig. The rapid flow of digesta in this region does not favour high levels of bacterial growth but the slower digesta movement in the ileum and large intestine permits bacterial concentrations of up to 10^{11}/g digesta. Many of the enzymes in or on bacteria operate slowly on complex, and often branched, polysaccharides, with this occurring in the distal part of the gut.

The specific bacterial carbohydrates found in animals have not been studied in detail, but many reports provide indirect evidence that they have a very wide range of activity because of the great variety of substrates that can be fermented, for example, by pigs. Most of the fermentation leads to production of lactic, acetic, propionic and butyric acids, with the evolution of carbon dioxide, water, hydrogen and methane. It has been widely assumed that the stochiometry of this activity in pigs follows the same principles as those which apply in the rumen, but recent studies have suggested that this is not entirely so because of the low levels of methane which have been found by Zhu *et al.* (1988).

The bacteria of the gut produce a wide range of enzymes and currently there is interest in the use of exogenous enzymes, produced by microbial fermentation. The genus *Bacillus*, for example, is well known for producing a wide variety of enzymes, including polysaccharases, proteases and nucleic acid-hydrolysing enzymes (Table 2.13). Amylase from *B. subtilis* was considered an industrial enzyme even by 1917.

Bacterial numbers in the gastro-intestinal tract

The total number of organisms within the tract per gram of contents also varies between different regions. The stomach generally contains the least (10^6–10^7) gram of contents while the colon contains the most (10^9–10^{10}) gram of contents.

CHAPTER 3

The microbial cell

Introduction

While the species which make up the flora and fauna of the gastro-intestinal tract are numerous, bacteria play a key role. Consequently, it is relevant to give particular consideration to the form and function of the bacterial cell.

The structure of the bacterial cell

Bacterial cells occur in a variety of shapes and sizes depending on the type of organism and on the way in which they have been grown (*Figure 3.1* illustrates the basic structure). They are smaller than animal cells and are more subject to osmotic and environmental changes. *Plates* 3.1 and 3.2 show photographs taken with an electron microscope of *Lactobacillus fermentum* and *Bacillus subtilis*, two typical non-pathogenic organisms.

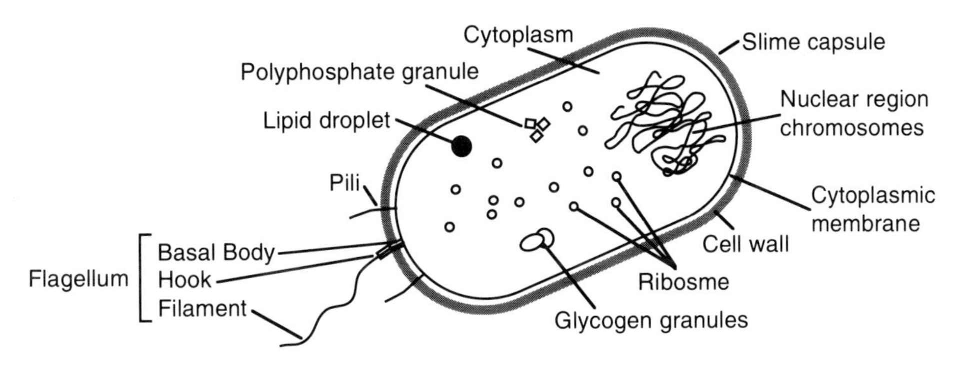

Figure 3.1 Simple illustration of a bacterial cell.

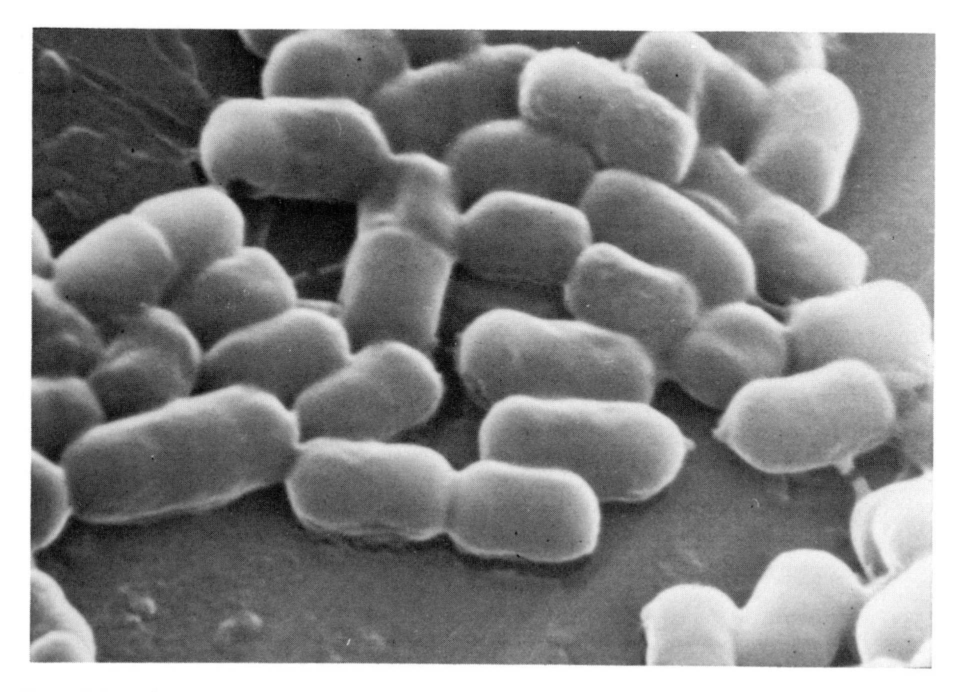

Plate 3.1 Electron microscope photograph of *Lactobacillus fermentum*

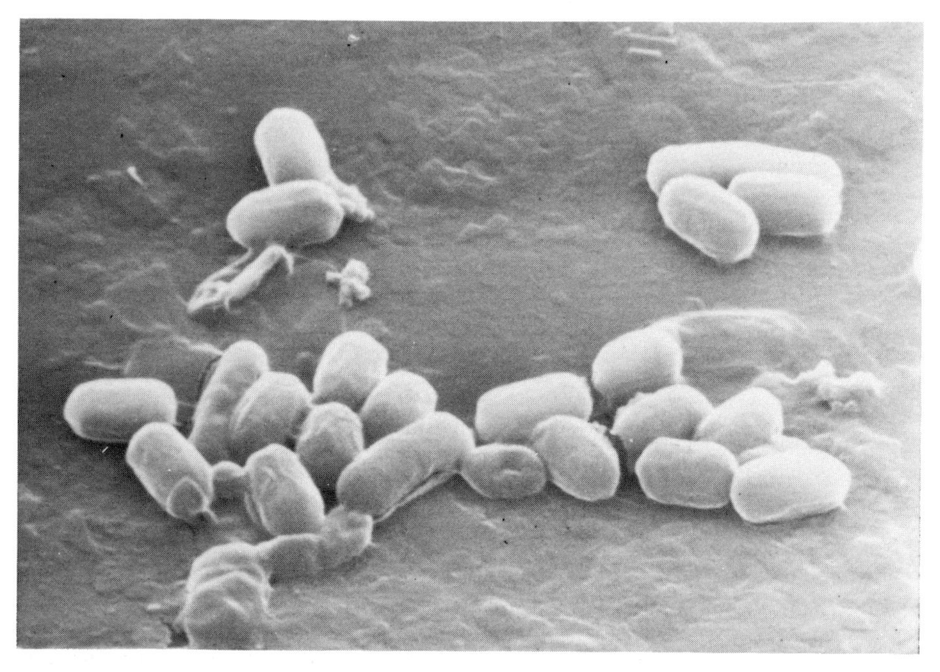

Plate 3.2 Electron microscope photograph of *Bacillus subtilis*

(In plate 3.2 some lysed cells can be seen). Bacteria can be spherical, curved or spiral; however, most are rod-shaped and are about 1 μm wide and 2 μm in length (1 μm = 0.001mm). A single bacterial cell has an approximate volume of 10^{-12}ml and contains 2.5×10^{-13}g dry matter. The bacterial cell is very organised, with DNA embedded in the cytoplasm surrounded by the cell membrane (called the protoplast), and outside this lies the cell wall.

Cell wall

Almost all bacteria possess a cell wall, a rigid outer structure which is the protective cover of the cell (*Figure 3.1*), but is itself the site of chemical action, e.g. the production of many antibacterial compounds. The cell wall structure can vary considerably between bacteria but the main difference which helps in classification is their reaction to the Gram stain. **Gram-positive bacteria** normally have a thick, relatively homogeneous cell wall (see *Figure* 3.2), and although different layers can frequently be distinguished by

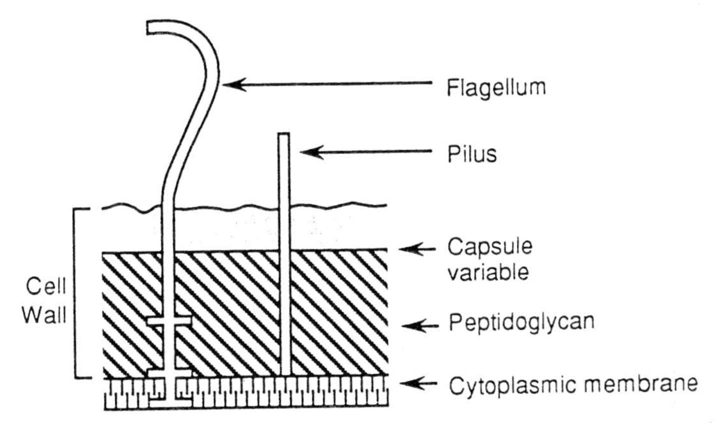

Figure 3.2 Cell wall structure and basal system of the flagellum in Gram-positive bacteria, e.g. *Lactobacilli*.

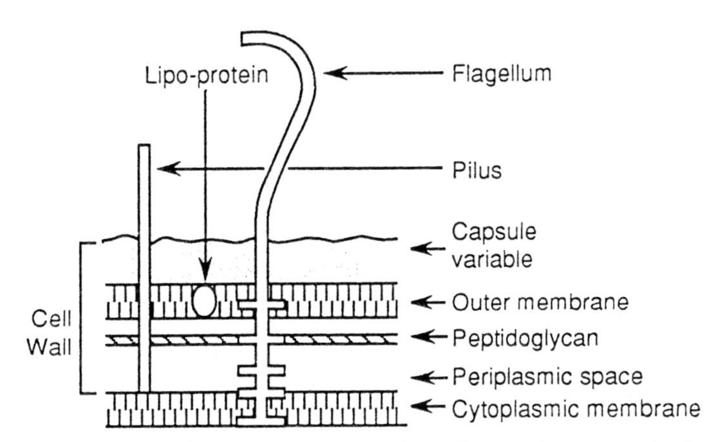

Figure 3.3 Cell wall structure and basal system of flagellum in Gram–negative bacteria, e.g. *E.coli*.

staining density, they are seldom sharply defined. In contrast, **Gram-negative bacteria** normally have a thinner, distinctly layered cell wall, which includes an outer membrane resembling the cytoplasmic membrane in structure (*Figure* 3.3). Gram-negative bacteria are those that cannot retain the complex of basic dye and iodine in the face of the decolouring agent, and have cell walls which are more porous than Gram-positive cells which resist de-colouring. Gram-negative bacteria (e.g., *E. coli*) are more adaptable to their growth conditions than Gram-positive (e.g., *B.subtilis*). They utilise inorganic nitrogen compounds, mineral salts and simple carbon sources for formation of the cell. Gram-positive bacteria tend to be more specific in their nutritional requirements, often lacking synthetic abilities, requiring some amino-acids, vitamins, and growth factors. Consequently they are usually grown in rich broths and are less likely to colonise in difficult conditions.

The mechanical strength of the cell wall is evident when, under strictly controlled conditions it is removed to release the remainder of the cell (a free protoplast) in an intact state. Regardless of the original shape of the cell, the protoplast is spherical and, without the protection of the cell wall, is osmotically sensitive. The cell wall is essential for cell division.

Gram-positive bacterial cell wall

The wall lies outside the cell membrane and is usually between 15 and 30nm thick (see *Figure 3.1*). Most of the cell wall is made up from two large polymers. Peptidoglycan forms over 50% of the wall weight and is a cross linked structure providing a strong multi-layered mesh (Nakagawa *et al.*, 1984) giving the cell shape and allowing it to withstand high internal osmotic pressure. The rest of the wall mass differs between species and can be made mainly from teichoic acid and N-acetylglucosamine 1–phosphate. The cell is strongly polar with a negative charge, with proteins and polysaccharides often found on the outside of the wall giving the bacteria antigenic properties. A small percentage of bacteria also have complex lipids in their walls.

Gram-negative bacterial cell wall

In these bacteria the wall is usually more complex than Gram-positive organisms. The cell wall appears as a structure containing three layers separated by clear layers (under electron microscopy). The clear layer which lies on the outside of the cytoplasmic membrane is termed the 'periplasmic space' (see *Figure* 3.3). It contains soluble enzymes and other compounds necessary for metabolism. Outside the periplasmic space is a layer of 3nm thickness made mainly of peptidoglycan (constituting 5% of cell weight compared with 50% in Gram-positive bacteria).

All the structures on the very outside of the Gram-negative cell wall together make up the 'outer membrane' (6–10nm in thickness). The outer membrane is made up of proteins, lipo-proteins, lipo-polysaccharides, phospholipids and fatty acids (Sutherland, 1985). Polysaccharides again make up the outside of the cell, giving it antigenic properties. The outside of the cell contains small pits of 2nm diameter made up of protein molecules called 'porins'.

There are 20–30 proteins in the outer membrane which alter the permeability of the cell. This makes it act as a molecular sieve with many molecules passing through by simple diffusion. However, many of these proteins in the outer membrane which act as specific pores, allow entry of molecules too large for the regular pores, e.g. Fe^{3+} complexed with ferrichrome and vitamin B_{12}.

Peptidoglycan is one of the main components of the cell wall and is said to make it rigid (*Figure* 3.2 *and* 3.3). The peptidoglycans comprise chains of amino sugars, N-acetylglucosamine alternating with N-acetylmuramic acid (*Figure* 3.4).

Peptidoglycan is not found in cells other than those of prokaryotes. The walls of eukaryotic micro-organisms, such as yeasts, maintain rigidity, etc., with a thick enmeshed (but uncross-linked) array of β–glucans, mannan-protein complexes and other largely carbohydrate polymers.

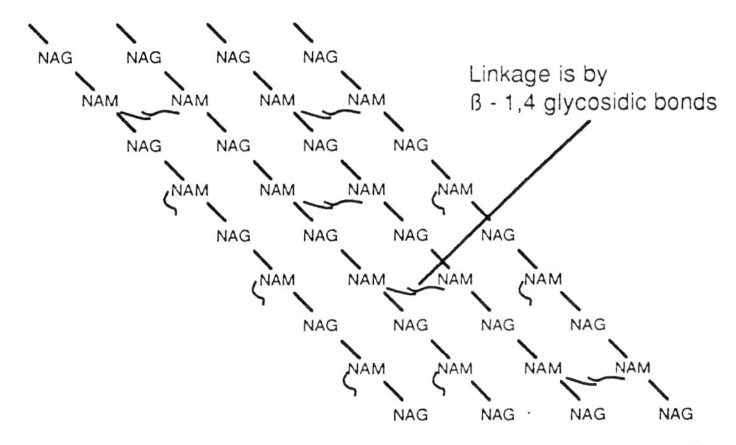

Figure 3.4 The basic structure of peptidoglycan in *E. coli*, a Gram-negative bacterium. NAG = N-acetyl glucosamine, NAM = N-acetyl muramic acid.

Teichoic acids: structure

Teichoic acids are polymers containing phosphates which occur probably in all Gram-positive bacteria. They appear to be absent from most Gram-negative species.

Teichoic acids play an essential role in the maintenance of adequate concentrations of metal ions. Mg^{2+}, in particular, is held in the area of the cytoplasmic membrane's outer surface. Most membrane functions and enzymes require significant Mg^{2+} concentrations.

It has been suggested that wall teichoic acids provide additional magnesium binding capacity for the surfaces of cells which can grow in the presence of high concentrations of NaCl. *B. subtilis*, grown in normal media, contains a wall teichoic acid.

Non-protein components are connected with polypeptide chains (*Figure* 3.5).

Figure 3.5 shows how non-protein components help connect polypeptide chains to form a rigid cell protein.

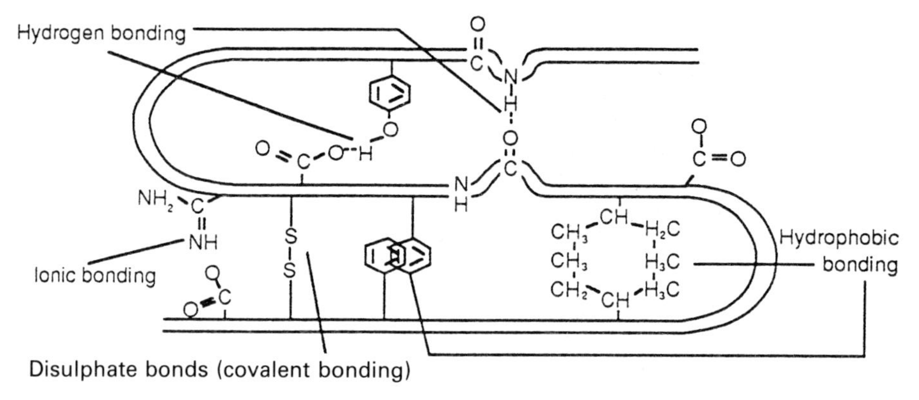

Figure 3.5 Intermolecular bonding of polypeptide chains in cell protein. (Based on Schlegel H.G., 1986)

Cytoplasmic membrane

The cytoplasmic membrane usually adheres closely to the cell wall and plays a vital part in the absorption of nutrients into, and the excretion of waste products from, the cell. It also contains many enzymes, some of which are associated with respiration whilst others are involved in the production of cell wall material.

Intracellular materials

Nuclear material

The nucleus of bacterial cells has the same function as that of nuclear materials in the cells of plants and animals. It acts as a control centre and passes on genetic instructions from the cell to its daughter cells. Unlike the nuclei of higher animals the nuclear material of bacteria does not possess a limiting membrane or nucleolus and divides by simple fusion. If cells are studied during a period of rapid growth, four or more nuclei may be seen in a single cell preceding the formation of cross-walls.

DNA

The primary structure of all the cell proteins are encoded in the genetic material, which is ordinarily DNA. Deoxyribonucleic acid (DNA) is a macromolecule (*Figure* 3.6), which upon acid hydrolysis yields equimolar amounts of its three components: deoxyribose, phosphoric acid, and nitrogenous base.

The DNA molecule contains four different bases: two purines (adenine and guanine) and two pyramidines (cytosine and thymine) (*Figure* 3.7).

Because DNA contains the genetic information of the cell, its replication always precedes cell division, and must lead to the formation of two completely identical DNA chromosomes. In addition to chromosomal DNA, many bacteria contain extra chromosomal DNA in closed, circular, double-stranded forms called plasmids.

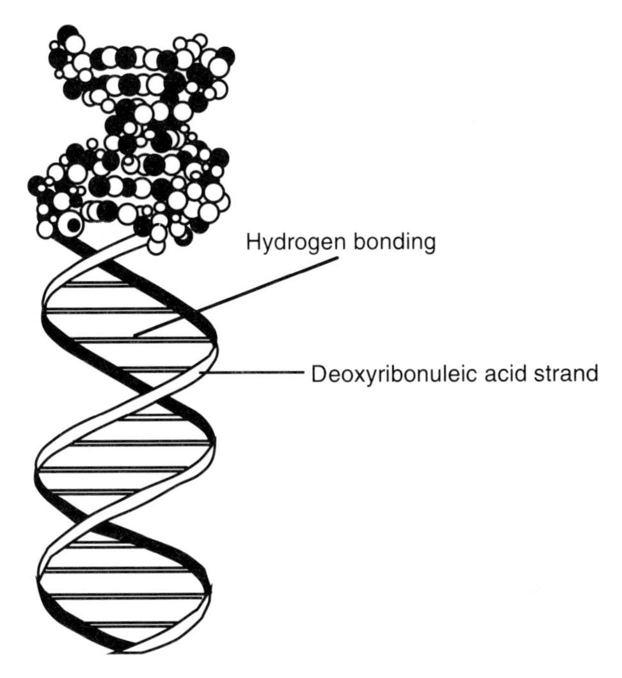

Figure 3.6 The structure of DNA. Two stands of DNA are joined by hydrogen bonds to form a double helix (Based on Schlegel H.G., 1986).

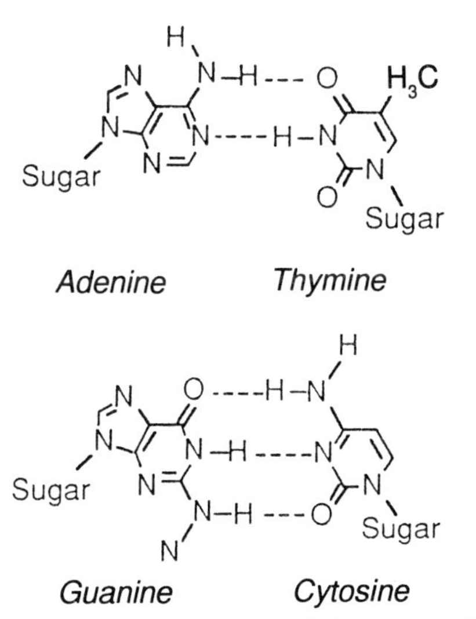

Figure 3.7 Hydrogen bonding between adenine and thymine, and cytosine and guanine in DNA (Based on Schlegel H.G., 1986).

DNA controls protein structure via messenger RNA, which is transcribed directly from it. The proteins have both structural and enzymic functions which determine the form and metabolism of the cell. The genetic material therefore determines the make up of the cell.

DNA and organism identification

The sequence of the four bases of DNA codes for the sequence of amino acids in a protein. This sequence of bases can be recognised by restriction enzymes. This forms the basis for establishing different bonding patterns in different organisms using Southern Blotting to show the differences as restriction fragment length polymorphisms (RFLP). Each organism has its own unique bonding pattern and DNA mapping or fingerprinting and is a useful tool in identification. For example, *Saccharomyces cerevisiae* is very important in animal nutrition and it is now recognised that different strains of yeasts have different properties (Moore, 1993). DNA fingerprinting is used to identify different strains (see *Plate* 3.3).

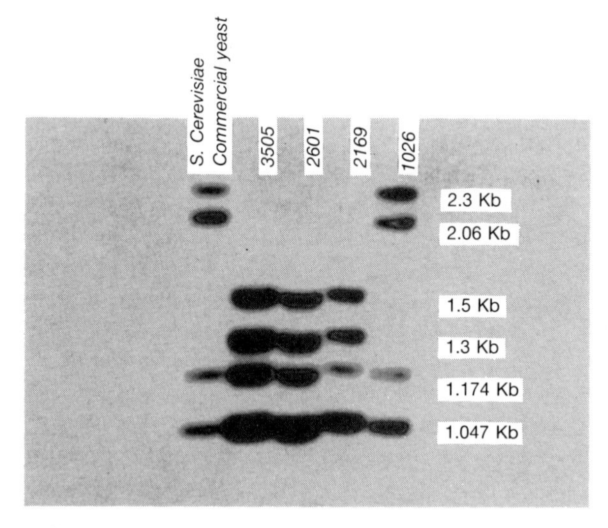

Plate 3.3 Genetic fingerprinting of various yeast strains. The pattern of the *Saccharomyces cerevisiae* 1026 is replicated by the commercial yeast 1026 and thus identifies it. [*By courtesy of Elizabeth Moore*]

DNA and genetic manipulation

In the mid-1970s the new science of biotechnology was given a boost when DNA cloning, oligonucleotide synthesis and gene expression were used together, for the first time, to express human protein from recombimant DNA, which involves genetic manipulation by combining DNA sequences from different sources.

This aspect of biotechnology presents many opportunities in agriculture and medicine from large livestock to micro-organisms. The latter are particularly suited to this technology and they have important roles in the production of many substances, e.g. enzymes. Consequently, the opportunity exists to produce organisms which are well suited to particular tasks.

Ribosomes

The cytoplasm of bacteria contains tens of thousands of minute granules called ribosomes. They comprise ribonucleo-protein, and a large part of the ribonucleic acid content of the cell. They can only be seen under an electron microscope and are involved in the production of proteins.

Other intracellular granules

Granules consisting largely of inorganic metaphosphate may be present in cells, and can be seen under a standard light microscope. They are known as volutin granules and can be seen to become larger and more abundant when the cell is in a favourable environment, and diminish in size when the environment is antagonistic. They appear to act as food reserves. Other granules are present, e.g. lipid granules and others composed of sulphur and glycogen, however, they probably have a similar function.

Cell compositon

The cell wall composition is dependent on growth rate, temperature, pH and the nature of the growth-limiting constituent. Most of these parameters, however, vary during normal growth and therefore, the cell wall composition also varies. In Gram-positive *Bacilli* the ratio of cell wall weight to total cell dry weight decreases as the cells grow faster. In contrast, little effect of growth rate is seen in *Staphylococci* or *Micrococci.*

Approximately half of the dry weight of bacteria is protein with more than 1000 different species of protein in any one cell. The molecular mass varies from a few thousands to millions, each having a very definite and precise composition with the sequence of amino acids giving the cell its primary structure. The cell is biologically active because of its precise three dimensional conformation and protein content, which has catalytic functions.

Extracellular structures

Capsules and loose slime

Many bacteria, Gram-positive and Gram-negative alike, secrete a slime covering or capsule which in most cases is polysaccharide (sugar and amino acids) (see *Figure 3.1*). Although its synthesis is optional, depending on the growth medium, and although growth occurs readily without capsule formation, this substance is frequently the major determinant of a bacterial cell's ability to colonise a given niche (such as *Streptococcus salivaricus* on teeth). Capsules are usually formed by certain pathogenic and saprophytic bacteria and are circumscribed gelatinous layers outside the cell wall. Pathogenic strains with a capsule are more virulent than non-capsulate strains. This increased virulence is associated with the ability of such strains to avoid or tolerate phagocytosis by body defence cells.

Capsular material consists mainly of water with a low proportion of solids. The solids are usually complex polysaccharides being highly specific both chemically and serologically.

Loose slime is formed by many capsulate and some non-capsulate species. In the former the slime is often very similar chemically and antigenically to the capsular material. When grown in fluid media the slime often disperses, but when grown on a suitable solid medium the secreted slime remains in association with the bacterial cells and the resulting colony acquires a mucous appearance and consistency.

Flagella

Many micro-organisms are motile by the use of simple strands of protein called flagella which are twisted spiral filaments and are usually much longer than the cell. A flagellum consists of three parts, the body, the hook and filament or shaft. The basal body is a complex structure which anchors the flagellum in the cell envelope (*Figures* 3.2 and 3.3). It is joined to the filament through the protein hook region. In some bacteria (eg. *Vibrio* species) the filament has an outer sheath which is continuous with the outer membrane of the cell wall.

Some motile bacteria (*Figure* 3.8) are:

- Monotrichous – only one flagellum and it is attached at one of the two poles.
- Amphitrichous – a single flagellum at both poles.
- Lopotrichous – two or more flagella extruding from one or both poles.
- Peritrichous – numerous flagella originating from the entire cell surface, e.g. *Salmonella*.

The type of flagella present determines the bacteria's ability to move in its environment.

Motility is achieved by the rotation of the flagellum from the basal body – i.e., the flagellum acts as a propeller to push the body through its medium (*Figure* 3.9). Flagellated bacteria are therefore more likely to colonise in areas of the digestive tract that non-flagellated bacteria could not reach. Flagella are slender, (usually 12–18mm in width), so they cannot be seen under the light microscope. In any one species the distribution of flagella is constant.

Fimbriae

Bacteria also possess structures on their surface that are shorter and more rigid than flagella and about half the width. They have no association with motility. These short, straight, and hair-like bacterial appendages are named fimbriae (Latin, threads) or pili

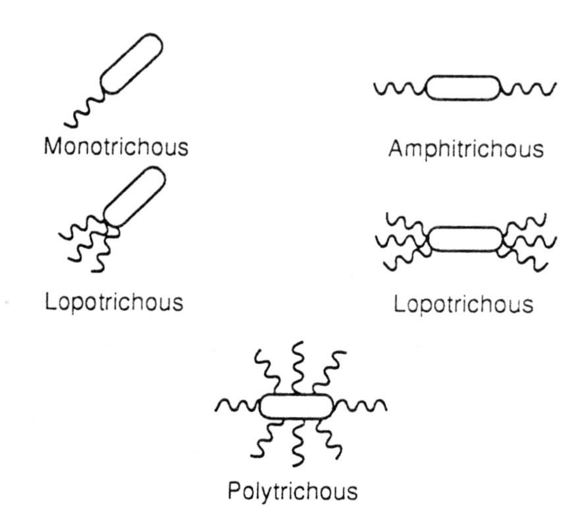

Figure 3.8 Bacterial flagella.

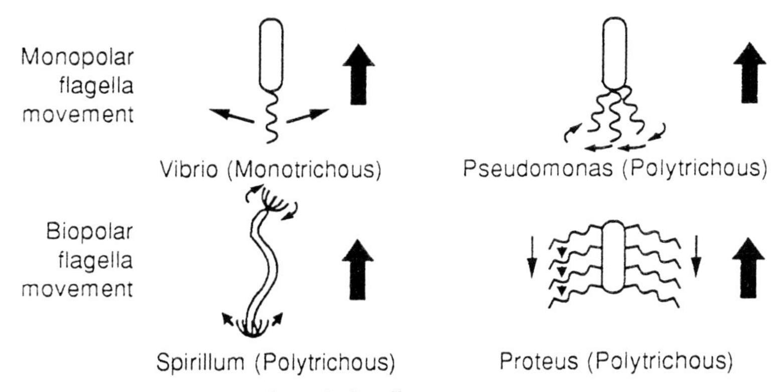

Figure 3.9 Bacterial movement through flagella.

(Latin, hairs). Fimbriae are made from a single protein and can extend 10 micrometres into their surrounding medium. They appear to be organelles of attachment showing specificity (Ingraham *et al.*, 1983). The lectins of the fimbriae recognise and attach to particular oligosacharrides of the gut wall. There is also good evidence that fimbriae of enteric bacteria cause aggregation of the bacteria to form a pellicle on the surface of static cultures. It appears that the function of the pellicle is to help fimbriated strains compete better for oxygen, which is in limited supply below the surface of the medium. *E. coli* possessing fimbriae form smaller more dense colonies on solid medium than non-fimbriated strains. Fimbriated cells grow more rapidly than non-fimbriated cells in liquid media when oxygen is limiting. In fact, the number of non-fimbriated cells decreases during the stationary phase of growth, while the number of fimbriated cells increases. This behaviour could be an advantage in the environment of the intestine where oxygen is usually limiting.

Cell growth

Cell growth is characterised by a number of phases (see *Figure* 9.1). Inoculation of a nutrient medium is followed by a **lag phase** which is a period of adaptation. When a microbial culture is shifted from one environment to another, it is necessary to re-organise both its micro- and macro-molecular constituents. This may involve the synthesis or repression of enzymes or structural components of the cell. As a consequence, the lag phase may be very short or quite lengthy. During this phase, cell mass may change without a change in cell number. There can also be an apparent or pseudo lag phase when the inoculum is small or has low viability, since growth can only be registered above the sensitivity level of the method of analysis. This is a period of balanced or steady state growth called the **log phase**. Throughout any fermentation the chemical composition of the surrounding environment is continually changing as nutrients are being consumed and metabolic products are being formed. The growing environment is therefore not in a steady state. During the lag phase, the cell macromolecular composition remains constant.

This **stationary phase** occurs when all the cells have stopped dividing or when viable cells have reached equilibrium with dead cells, that is, with the rate of death (see *Figure 9.1*). Even though net growth has stopped, two things may happen. (1) There may still be metabolism and accumulation of products in the cell or in the surrounding environment. The total cell mass may stay the same but the number of viable cells will undoubtedly fall. (2) Viability may decrease as cell lysis occurs and cell mass falls. This leads to the creation of a complex medium of lysis products, and often to a secondary period of growth called cryptic growth. Further details on cell growth can be found in Chapter 9.

Types of microbial growth

Bacteria divide by fission (*Figure 3.10*). During the growth cycle the cell doubles its mass and the amount of all cell constituents. Prior to fission, the cell wall and membrane are synthesised and the parent cell divides into two identical daughter cells, each of which grows exactly as the original cell. It is generally not possible to determine the age of a bacterium except within the context of a single cell doubling.

Cell division differs in different organisms. Examples of bacteria, yeast and fungal mycelia are given in *Figure 3.11*.

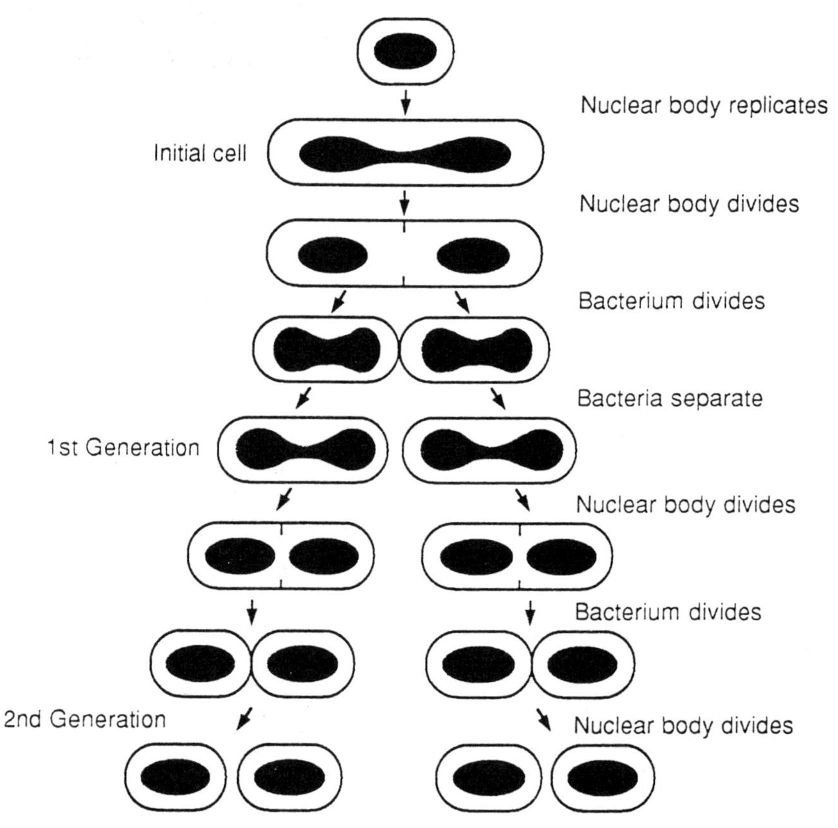

Figure 3.10 Cell division.

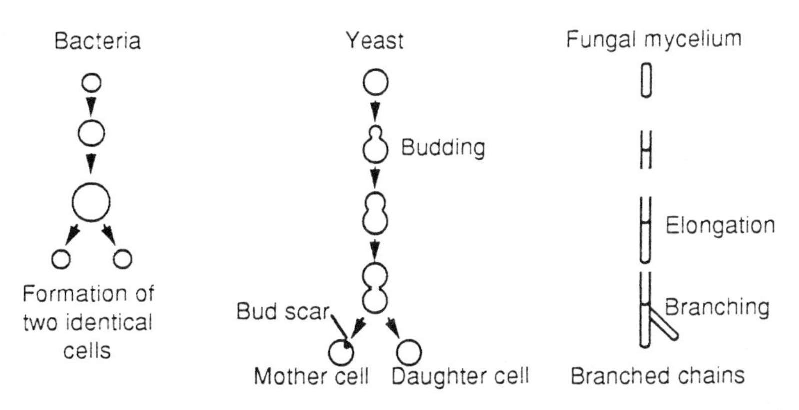

Figure 3.11 Types of cell division seen in different microbes, i.e. bacteria, yeast and fungal mycellium.

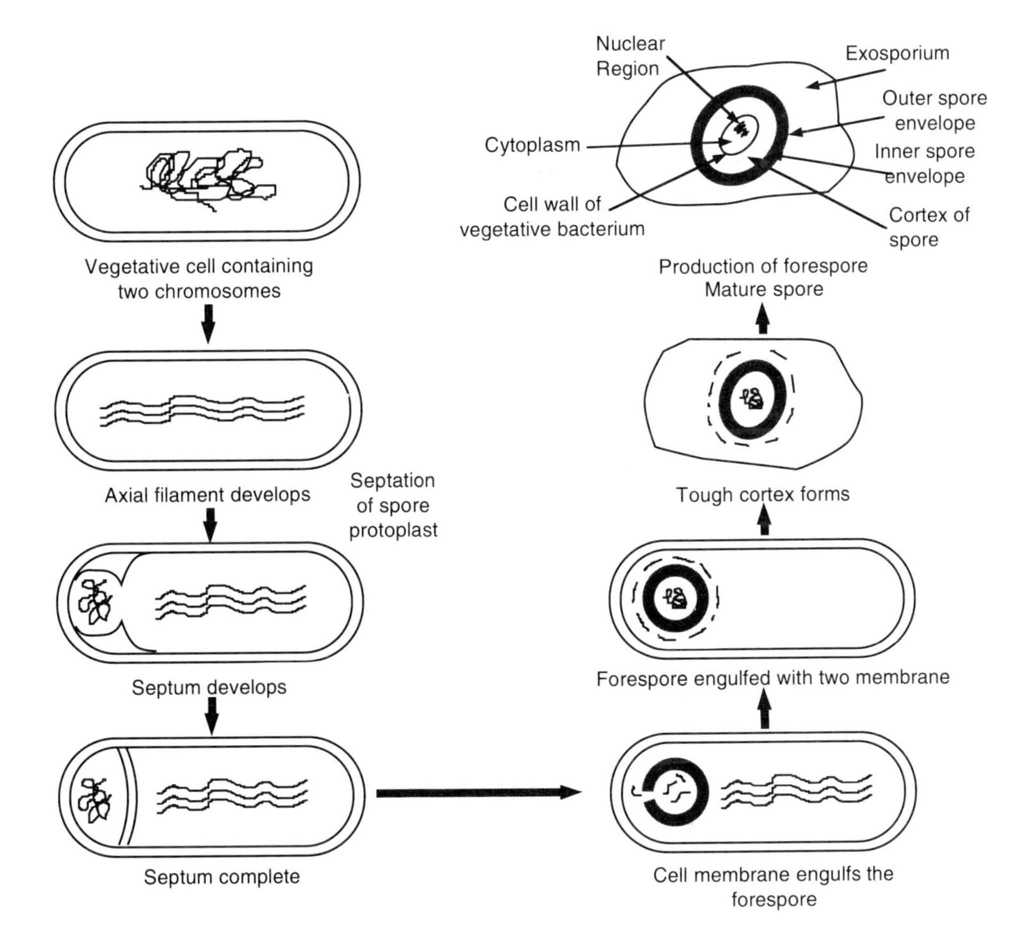

Figure 3.12 Formation of spores.

Micro-organisms that require oxygen for their energy yielding pathways are called aerobes, while those that cannot utilise oxygen are called anaerobes. Organisms that are capable of using either respiratory or fermentation processes depending on the oxygen supply of the environment are called facultative.

Bacterial endospores

These are usually called spores and are extremely resistant to adverse environmental circumstances which may be lethal to the vegetative cells which produce them. Their resistance to extreme conditions is due to a hard spore case, a low water content and extremely low metabolic and enzymic activity.

As only one spore is formed from a vegetative cell and on germination of the spore only a single vegetative cell develops, they have no reproductive significance (see *Figure 3.12*). Spores are generated from *Bacillus* and *Clostridial* bacteria.

Fungi

Thousands of species of fungi are known to man, with yeasts and moulds only seen under the light microscope. Larger species such as mushrooms can be substantially larger. *Figure 3.13* shows the fungal classification.

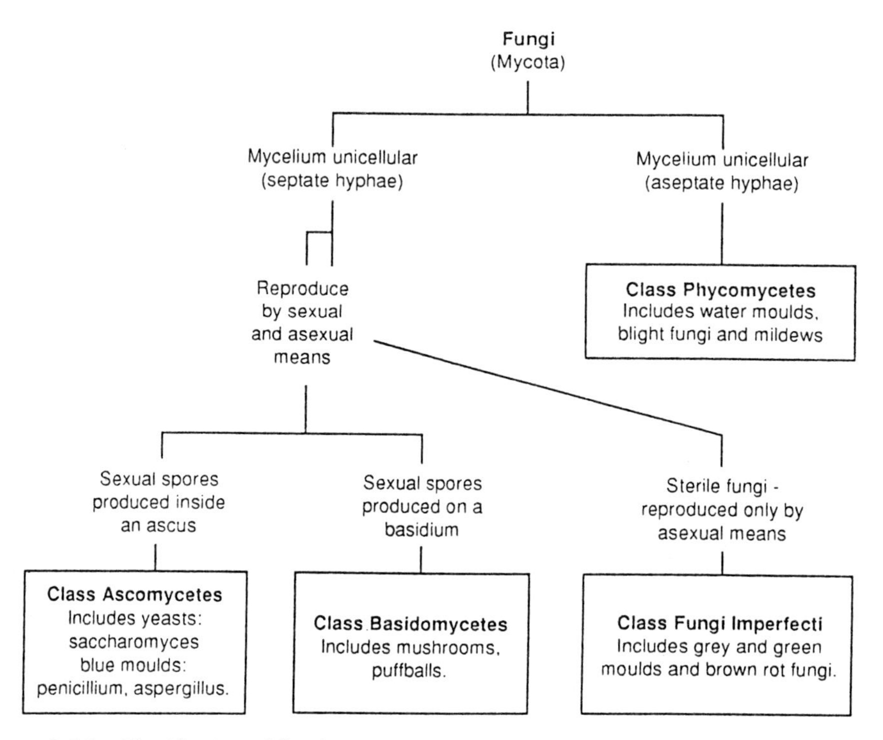

Figure 3.13 Classification of fungi.

Plate 3.4 Budding yeast cells *(Saccharomyces cerevisiae). [By courtesy of Kirk Robinson]*

Fungi have over the years caused death and disease in living organisms. Some species affect crop yields while others affect man and domesticated animals, e.g. athletes foot and ringworm. However, their biochemical activity has also been long exploited by man. Yeasts have been used in bread, beer and wine-making, while in the 1920s Fleming discovered penicillin from a mould.

Fungi are usually formed from entangled hyphae (branched treads) forming a mass known as mycelium. The hypha may be 0.5 μm to 1.0mm in diameter. Aseptate hyphae are single multi nucleate protoplasts in the cell wall. Septate hyphae comprise protoplasts with one or several nuclei. The smallest fungi include yeast (*Saccharomyces*) (see *Plate 3.4*) which is unicellular (see *Figure 3.14*).

Yeast has a distinct nucleus surrounded by a nuclear membrane (which is porous). Chromosomes and a spindle appear during cell division.

The cytoplasm contains many organelles, such as mitochondria, and extensive endoplasmic reticulum which are not found in bacterial cells. Ribosomes and many vacuoles contain storage media such as glycogen, metaphosphate polymers, lipid globules and volutin. The protoplast of the yeast cell is enclosed by plasma membrane, which is in itself enclosed in a rigid, permeable wall. This cell wall of yeast is composed largely of polymers of mannose (mannans) and glucose (glucans), but unlike other fungi is not made up largely of chitin (a polymer of N-acetyl glucosamine).

The filamentous fungi are quite similar in structure to yeasts with the young tips of the hyphae containing ribosomes, mitochondria, and small vesicles packed in the

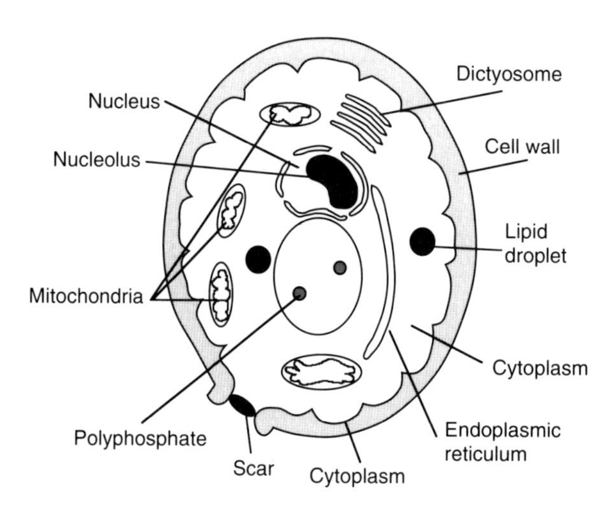

Figure 3.14 Cross section of a yeast cell.

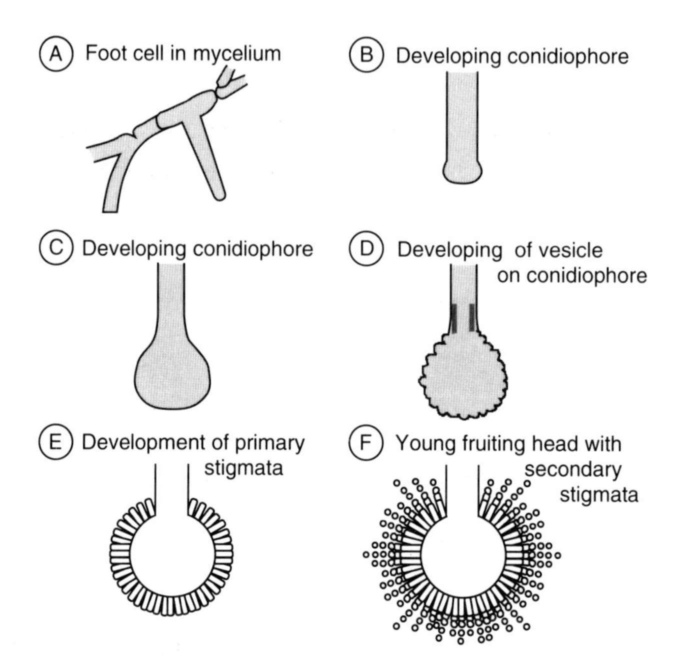

Figure 3.15 Development of *Aspergillus* from immature foot cells to mature fruit heads.

protoplasm. This tip is the site of synthesis of new wall material, organelles and essential enzymes. In addition to yeast (*Saccharomyces*) animal nutrition has also focussed on the fungus Aspergillus. It is a member of the *Ascomycetes*. Unlike other fungi the most common asexual reproductive unit is the conidiophore. The spores are formed inside phialides, while the conidophores are more complex than in other fungi (see *Figure 3.15*)

CHAPTER 4

Micro-flora of the gastro-intestinal tract

Introduction

The digestive tract is colonised soon after birth by a variety of micro-organisms. The first bacteria in the intestinal tract, including the *Lactobacilli*, presumably originate from the mother's vagina during birth or from the surrounding environment. *Lactobacilli* constitute an important part of the vaginal micro-flora, especially in pregnant women with the normal vaginal micro-flora apparently dependent on the glycogen content of the vaginal epithelium which is itself dependent on ovarian activity (Speck, 1976).
These *Lactobacilli*, as well as *Streptococci* and Enterobacteria, are out-numbered in the gut of poultry and mammals by strictly anaerobic bacteria (Hentges, 1983; Hungate, 1966; Savage, 1986; Steffen and Berg, 1983; Stephen and Cummings, 1980; Wolin, 1981). The bacteria of the gut are affected by the host's nutrition (Anderson and Pepper, 1963 and Anderson *et al.*, 1953) and the environment of the animal (Rosebury, 1962; Dubos *et al.*, 1965).
Micro-organisms associated with the intestinal tract have been divided into two groups.

- **The autochthonous micro-organisms**. These are indigenous and colonize a particular region of the intestinal tract early in life. They multiply to high levels soon after colonisation, and remain so throughout the lives of healthy, well-nourished hosts.
- **The non-autochthonous micro-organisms**. These are the indigenous organisms which colonise the intestinal tract of animals living in a given area but may not be in all individuals of a given species (Savage, 1972).
  Habitats can be divided into three, namely, luminal, epithelial and cryptal (*Figure 4.1*).

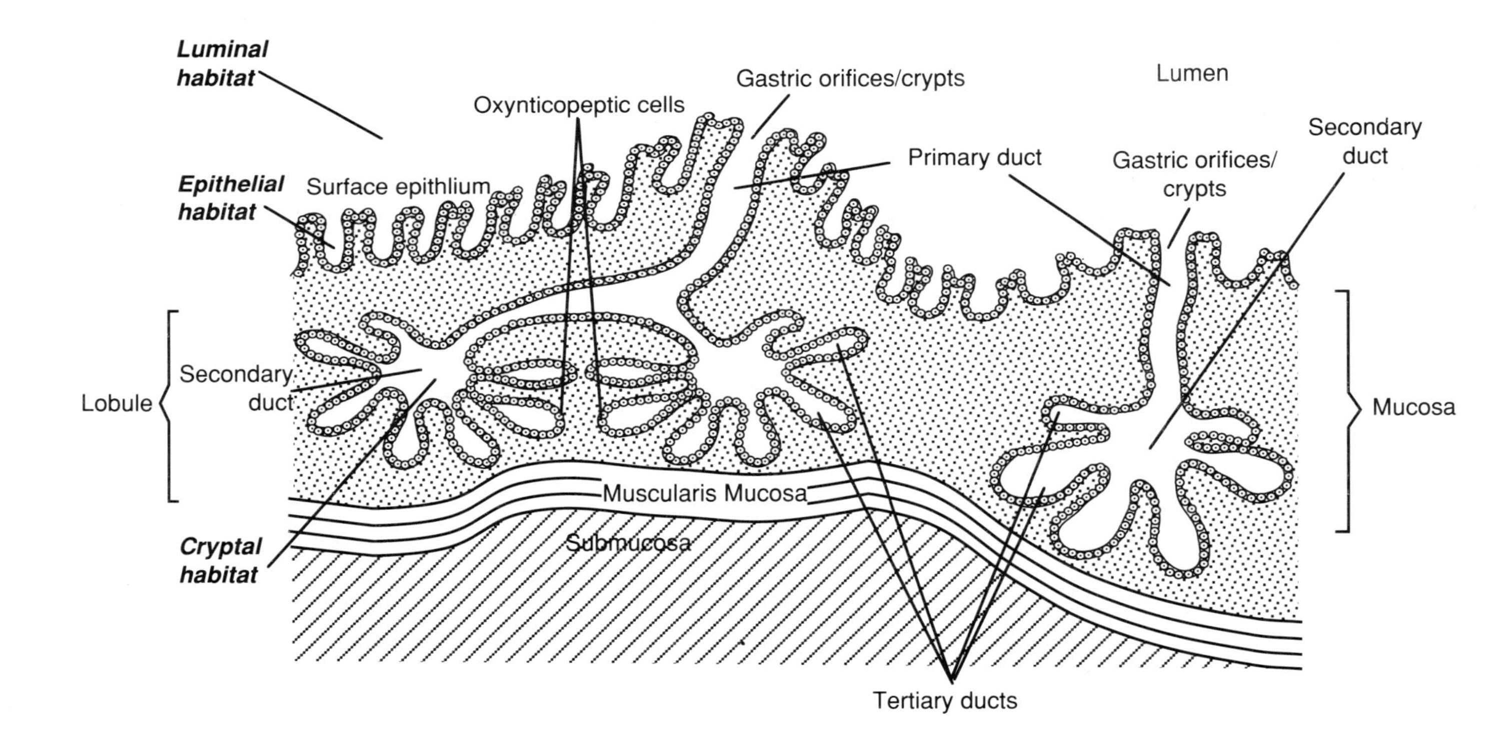

Figure 4.1 Illustration of mucosa (proventricular) with lobular glands present.

Certain types live free in the lumen (Rosebury, 1962), while some grow on the surfaces (Savage, 1972a); others colonise the crypts of Lieberkuhn and the epithelium (Savage, 1977, 1983, 1984, 1985, 1986, 1987, 1989). For example, in mice and rats, the *Lactobacilli* are found among the non-secreting gastric epithelia and are also attached to the digesta (Savage, 1989). It should be noted that certain bacteria are specific for a given host species throughout life and are independent of environmental factors.

Within the gastro-intestinal population there are about 400 different types of bacteria in equilibrium with each other and with the host animal. There is constantly a selection of bacteria which can grow and colonise the gut to produce a micro-flora characteristic of each host species. Part of this selection is chemical, due to inhibitory agents like volatile fatty acids, hydrogen sulphide, bile acids, lysozyme and lysolecithin (Fuller, 1984) and immunoglobulins. When the bacteria overcome the inhibitory effects, they must then contend with the constant flushing effect of the peristaltic movement of food from the anterior regions of the gut. Bacteria remain in the gut in two ways:

- By attachment to the epithelial cells lining the intestine.
- By growing at a rate faster than the rate at which they are being removed by peristalsis. (Fuller and Cole, 1988).

The number of bacteria does, however, vary in the different sections of the gastro-intestinal tract. (Table 4.1).

Table 4.1 *Typical numbers (log_{10}/g fresh weight contents) of micro-organisms in various sections of the gastro-intestinal tract of pigs.*

	Stomach	*Duodenum*	*Ileum*	*Caecum*	*Rectum*
Lactobacilli	7–8	6–7	7–8	8–9	7–8
Coliforms	5–6	4–5	6–7	7–8	8
Clostridia	Nil	Nil	0–4	4	5
Enterococci	Nil	Nil	0–4	5–6	5–6

(Henderickx, Decuypere and Vervaeke, 1976).

The bacteria in the mouth probably originate from the tooth surfaces and soft tissues as well as from food entering the mouth (Gibbons and Van Houte, 1971; Rosebury, 1962). The stomach contents are acidic and therefore contain few bacteria (Waites, 1991), but often have yeasts (Savage 1983, 1986, 1987, 1989). However, spirochaetes and helical shaped bacteria are present (Fox and Lee, 1989; Kasai and Kobayaski, 1919; Lee and Hazell, 1988; and Lee *et al.*, 1988). *Bifidobacteria* are also found in the stomach of young suckling animals (Henteges *et al.*, 1984; Neut *et al.*, 1987).

The bacterial flora of the gastro-intestinal tract

The different bacteria have specific niches within the gut. For example, Smith (1965) and Fuller and Turvey (1971) stated that in the chicken crop, *Lactobacilli* predominate over coliforms. These *Lactobacilli* not only influence the crop but also the small intestine and form a symbiotic relationship with the host animal (Dubos *et al.*, 1965). The chicken crop is also home to *Streptococci* although they are in smaller numbers than the *Lactobacilli* (Smith, 1965). In addition to the species listed in Table 4.2, various spirochaetes, mycobacteria, fusiforms, *Staphylococci* and *Streptococci* have been cultured from the intestine.

Rabbits were unique among the young of all the species examined by Smith (1971) in that their stomachs and small intestines contained very few bacteria. This was due to the formation of antimicrobial substances in the stomach contents later identified by Rodriquez and Smith as decanoic and octanoic acids (Smith 1971). In normal situations, man also has few bacteria present in the stomach.

There is epithelial localisation of *Lactobacilli* and yeast in the rodent stomach (Brownlee and Moss, 1961; Savage and Dubos, 1967; Savage *et al.*, 1968). Yeasts

Table 4.2 *Common bacteria in the digestive tract of pigs and poultry.*

Pigs	*Poultry*
Bacteroides ruminicola	*Bacteroides spp*
Bacteriodes uniformis	*Bacteriodes fragilis*
Bacteriodes succinogenes	*Bifidobacterium bifidum*
Butyrivibrio fibrisolvens	*Clostridium perfringens*
Clostridium perfringens	*Clostridium beijerinckii*
Escherichia coli	*Clostridium spp*
Eubacterium aerofaciens	*Eubacterium spp*
Lactobacillus acidophilus	*Fusobacterium spp*
Lactobacillus brevis	*Gemmiger formicilis*
Lactobacillus cellobiosus	*Lactobacillus acidophilus*
Lactobacillus fermentum	*Lactobacillus fermentum*
Lactobacilus salivarius	*Lactobacillus salivarius*
Peptostreptococcus productus	*Micrococcus spp*
Proteus spp	*Streptococcus faecium*
Ruminococcus flavefaciens	*Streptococcus faecalis*
Selenomonas ruminantium	*Ruminococcus obeum*
Streptococcus bovis	
Streptococcus equinus	
Streptococcus faecalis	
Streptococcus salivarius	
Veillonella spp	

Adapted from Harrison and Hansen (1950); Smith (1965); Barnes *et al.* (1972); Tannock and Smith, (1970); Fuller and Brooker (1974); Salanitro *et al.*, (1974a), Fuller *et al.* (1978); Russell, (1979); Robinson *et al.* (1984); Varel *et al.* (1987) and cited by Wu (1987).

of the genus *Torulopsis* grow in layers in the mucin on the surface and also deep in the gland pits of the secreting epithelium of the stomachs of mice (Savage and Dubos, 1967). *Lactobacilli* also grow in layers on the keratinized squamous epithelium of the non-secreting portion of the stomachs of rats (Brownlee and Moss, 1961) and mice, (Savage *et al.*, 1968) from both specific pathogen-free and conventionally housed colonies. The two microbial layers are reported to be mutually exclusive (Savage and Dubos, 1967).

The bacterial flora takes some time to develop but is fairly stable in the adult animal. (*Figure* 4.2).

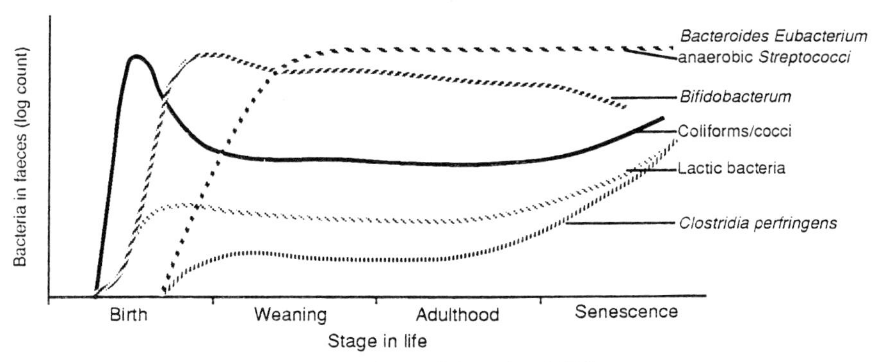

Figure 4.2 Intestinal flora changing with age (Mitsuoka, 1974)

Bacterial types

The population of bacteria in the gastro-intestinal tract is enormous (Table 4.1), since anaerobic and obligate anaerobic bacteria can be present at levels of 10^9 and 10^{11}/g of fresh digesta respectively (Ledinek, 1970; Savage, 1977). The large intestine allows the growth of bacteria by having a large volume of digesta coupled with a slow transit time of 20–38 hours (Mason, 1980). It has been estimated that there are 10^{14} bacteria in the gut and only 10^{13} eukaryotic cells in the body; thus the bacteria of the digestive tract are likely to have a major influence on the host's metabolism, physiology or nutrition (Luckey, 1963; Fuller and Cole, 1988). This high population of bacteria develops quickly after birth. High levels of *Lactobacilli* (10^4/g faeces) have been found only four hours post-partum. Twenty four hours later the level was 10^8–10^9/g (Muralidhara *et al.*, 1977). The acidity of the different regions of the gut affects the different types of bacteria that can survive there. In general, *E. coli* pathogens can survive at much lower pH than many non-pathogens.

Establishment of the intestinal micro-flora

Just before birth the digestive tract of all animals is bacteria free (Kenworthy and Crab, 1963). However, the gut flora develops very rapidly and within 2 hours of birth *E. coli* and *Streptococci* may be detected in the faeces of the newborn. This rapid colonisation of the gut can be complete within 2 days. The perinatal mammal encounters its first bacteria in the vagina of the mother, and then, after expulsion from the vagina, comes into contact with the large population of microbes present in both the faecal matter of its mother and within its new environment. By 5–6 hours after birth, the population of these two bacterial species is very high (10^9 and 10^{10}/g faeces, respectively). The rate of increase of bacteria varies from one species to another. For example, in one study with pigs the number of *Clostridium perfringens* was reported to increase rapidly after birth, whereas *Lactobacilli* appeared more slowly and only constituted the dominant portion of the flora 48 hours after birth (Ducluzuea, 1985). It was further observed that *Lactobacilli* and *E. coli* were present in the small intestine 4 and 8 hours respectively, after birth (Muralidhara *et al.*, 1972). *E. coli, Streptococci* and *Lactobacilli* have also been found in large numbers 18 hours after birth.

It appears that the new born animal possesses a very efficient microbial selection system, as it is able to modify the various bacterial ecosystems it encounters to produce relatively environmentally constant internal gut conditions. As already indicated, the first bacteria which become established in the digestive tract originate from the dam or the environment but they are not necessarily the most abundant ones of the ecosystems encountered by the young animal. For example, in a split litter experiment the five piglets obtained aseptically and kept in a sterile cage had a similar population of micro-flora of the small intestine at one week, to those reared on the sow (Ducluzeau *et al.*, 1975). There was a rapid establishment and implantation of *E. coli* and *Streptococcus* followed by *Lactobacillus* and *Clostridium*. However, clostridia were among the dominant species in the faeces or on the teats of the sow but failed to become dominant in the gastro-intestinal tract of the piglets reared with their dam. These studies were probably carried out under conditions which did not allow the detection of strict anaerobes, and if repeated today may have detected a wider range of bacteria.

Diet is important in the establishment of the gut micro-flora and that of the suckling piglet is interesting. While it may be assumed generally that it comprises mainly sow's milk together with supplementary solid feed, the piglets do consume considerable quantities of sow's faeces and bedding (Table 4.3). The nature of the faeces, particularly in relation to any harmful micro-organisms will be of great importance.

After initial establishment of the gut flora the number of bacteria declines, with the exception of *Lactobacilli* which become predominant; only in the large intestine do they co-exist with bacteroides and other strict anaerobes. This predominance of *Lactobacilli* is now generally recognised as desirable and is established when the newborn pig is

Table 4.3 *The typical diet of a suckling pig (g/day).*

	5 day old	*14 day old*
Milk (dry matter)	125	175
Supplementary feed (dry matter)	0	0–25
Faeces and bedding (fresh weight)	18	25

being suckled by its mother, but the mechanism by which the lactic acid producing flora suppresses the rest of the bacterial flora is far from understood. Some strains of *Lactobacilli* are known to colonise in thick layers the keratinized stratified squamous epithelium of the stomach of mice, rats and pigs, and the crop of chickens. (Wheater *et al.*, 1952; Fuller and Brooker, 1974). Under normal conditions, bacteria colonise according to a sequence characteristic of that particular part of the tract. (Mushin and Dubos, 1965; Schaedler *et al.*, 1965). The gut is thought to have a specific defence system against unwanted bacteria which may involve immunity as well as bacterial antagonism. Failure of this regulatory system leads the animal to possible invasion by enteropathogens (Tannock and Savage, 1974). Diet, drug administration, stress and the environment of the animal can also influence bacterial type (Tannock and Savage, 1974).

Method of bacterial selection of gastro-intestinal adhesion

The method of microbial selection determining the bacterial type has not been fully established, although many factors have been postulated to be involved. It appears that the diet may have some effect, since it is almost impossible to prevent neonatal *E. coli* diarrhoea in piglets that have not obtained any maternal colostrum. Maternal milk possesses bacteriostatic and anti-adhesive factors specifically directed against pathogenic *E. coli* (Chidlow, 1979). A possible role of lysozymes as antiseptic agents in sow's milk has also been suggested by Ducluzeau (1985). Among the environmental factors which influence the gut micro-flora, temperature may play a part because the number of coliform and clostridial bacteria have been reported to be slightly higher in the digestive tract of animals that had been kept in a cold environment. It has also been suggested that *Lactobacilli* and *Streptococci* which do not become rapidly established in the digestive tract may nevertheless exert a barrier effect towards the establishment of pathogenic *E. coli* (Ducluzeau, 1985; and Muralidhara *et al.*, 1977) but according to others, this situation has no effect on the gut flora (Fevrier *et al.*, 1979).

Specificity of colonisation of the gastric epithelium by *Lactobacilli*

Various strains of *Lactobacillus* are known to associate with the surfaces of the stratified squamous epithelium in the *pars oesophagia*, for example, of pigs (Fuller *et al.*, 1981) and in associating with the epithelial cells, the bacteria undoubtedly adhere to them (Savage 1981; Gibbons and Van Houte, 1971). Attachment of micro-organisms to the gut wall is by means of fimbriae (often referred to as ligands or pili). The bacterial fimbriae are lectins which recognise specific oligosacharrides which make up the attachment sites on the gut wall (*Figure* 4.3). The extent of colonisation does, however, vary both between the different bacteria and between animal species. Even the different regions of the intestine have different levels of colonisation (Barrow *et al.*, 1977). When the organisms have adhered, they multiply according to nutritional conditions prevailing in that area.

Many experiments designed to follow microbial colonisation of the gut have concentrated on studies with neonatal piglets and gnotobiotic chickens. However, in commercial production, treated animals would not be born into a germ-free environment and this difference may affect colonisation.

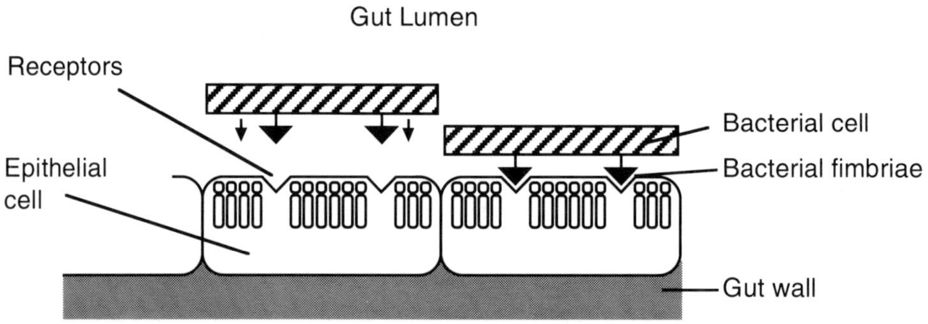

Gut Lumen
Receptors
Epithelial cell
Bacterial cell
Bacterial fimbriae
Gut wall

Figure 4.3 The frimbriae of bacteria recognise the oligosaccharide receptors on the gastro-intestinal epithelial surface. If the correct receptors are present colonisation occurs.

Stages in colonisation

In the first day or so of life, the whole alimentary tract is flooded with bacteria (i.e. pre acid stomach) (Smith and Jones, 1963), with the pH of the food helping in multiplication of ingested bacteria, which then pass unharmed into the small intestine and continue to multiply there. The lower pH of the stomach contents after the first day of life is usually enough to reduce multiplication of all ingested bacteria except the *Lactobacilli*. In pigs, *Lactobacilli* continue to proliferate in the stomach at all ages and, as a result, are the principal bacterial inhabitant of all parts of the alimentary tract except in younger piglets, where *Bacteriodes* in the caecum, large intestine and faeces are predominant. A sharp drop in pH occurs after a few days of life, which markedly suppresses the multiplication of these bacteria but not the *Lactobacilli*. A high pH of the stomach contents on the first day of life may result in clinical infection with enteropathogenic *E. coli* (Smith, 1971). *Figure 4.4* illustrates work of Barnes *et al.*, 1975, in chickens, showing the change in volatile fatty acid level and pH after hatching.

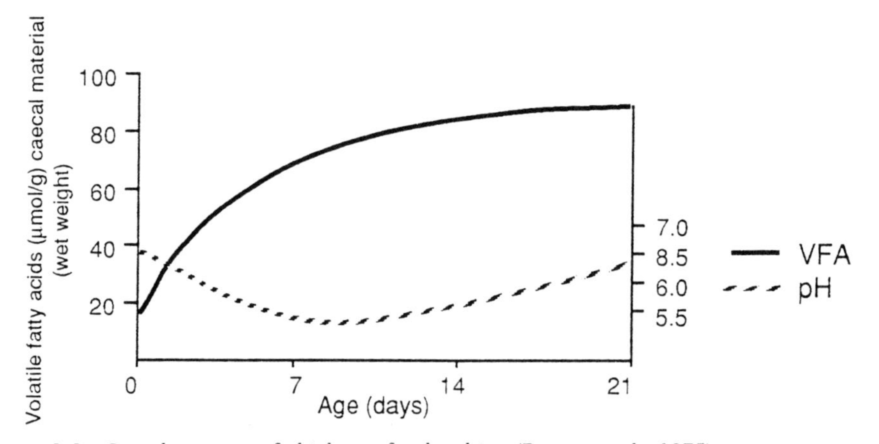

Figure 4.4 Caecal contents of chickens after hatching (Barnes *et al.*, 1975)
NB The volatile fatty acid (VFA) concentration and pH change up to 21 days of age. The pH rises after 7 days as mucosal buffering starts.

Adhesion of bacteria

The ability to attach to the gut epithelium is, for many organisms, essential in a continually transient environment like the gut (Gibbons and Van Houte, 1971). In this way they can avoid being swept away by the peristaltic flow of the digesta, and remain to inoculate fresh digesta as it arrives at the site of attachment. Thus, one method of preventing colonisation of the gut by pathogens is to saturate the adhesion receptors on the epithelium and prevent the pathogens attaching. This is one way in which live non-pathogenic bacterial supplements have been claimed to protect against intestinal disease. This epithelial cell adhesion is thought by some to be a host specific phenomenon and strains which adhere to pig cells will not necessarily adhere to chicken cells and vice versa (Barrow *et al.*, 1980; Kotarski and Savage, 1979). However, there may be some overlapping of adhesive powers.

Different bacteria have different adhesion determinants. Many different mechanisms of attachment have been described; for example, M-proteins for attachment of *Streptococcus pyogenes* to tonsils (Ellen and Gibbons, 1972). Adhesion of *Lactobacilli* to pig enterocytes has been reported to be controlled by the glycocalyx on the surface (Fuller and Brooker, 1980; Costerton *et al.*, 1985) and also the cell wall proteins (Wadstrom *et al.*, 1987). Adhesion ability also varies among lactic acid bacteria (Barrow *et al.*, 1980). Therefore, in theory, only lactic acid bacteria with high adhesive qualities, e.g. *Lactobacillus fermentum* and *Streptococcus salivarius*, should be selected for use in microbial additives. *Figure 4.5* illustrates how the correct cell ligand must be present to match up with receptors on the epithelial surface. Other factors can reduce adhesion, e.g. the presence of toxins and viruses (*Figure 4.6*). Adhesion of bacteria to components of food-stuffs is also common. *Plate 4.1* and 4.2 show lactic acid bacteria adhered to a starch granule.

Factors governing bacterial attachment

The bacterial flora which colonises the mucosa of the gastro-intestinal tract is controlled by the adhesion of bacteria to the epithelial cells, with colonisation of the mucopoly-saccharide component of the glycocalyx above the mucosal surface (Gaastra and DeGraaf, 1982; and Rozee *et al.*, 1982). It should be noted that readily fermentable carbohydrates are necessary for successful implantation (Hawley *et al.*, 1959).

A microcapsular layer on the surface of gastro-intestinal cells comprises mainly acidic carbohydrates and glycoproteins. These compounds have been associated with bacterial

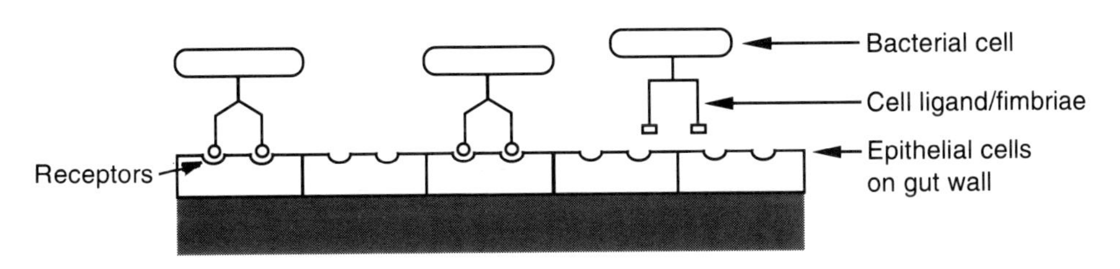

Figure 4.5 A bacterium with fimbriae which cannot be recognised by the receptors which cannot colonise the epithelial cells.

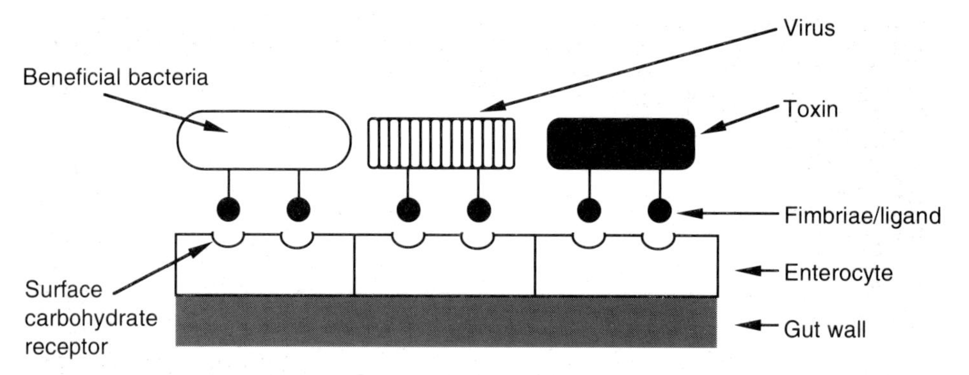

Figure 4.6 Viruses and toxins can link to the oligosaccharide receptors in the gut wall blocking attachment of beneficial organisms.

adhesion as attachment does not occur if the carbohydrate is neutralised or if glycoprotein is degraded with pepsin. The microcapsule of the bacteria and the glycocalyx of the epithelial cells are held close by electro-static attraction, with mucin reportedly needed for attachment of *Lactobacilli* and *Streptococci* (Reiter, 1979). Attachment is also helped by filamentous projections extending from the surface of the bacterium to the epithelial cells (Fuller and Brooker, 1974). The attachment of the bacteria allows multiplication to occur, producing high numbers of organisms within the gastro-intestinal tract. The organisms which are lost during the regular sloughing of the mucosa caused by the actions of peristalsis, are replaced by this production. The same type of epithelium from different host animals may possess different receptors, making it host specific (Van Houte *et al.*, 1972).

Other factors specifically affecting adhesion of *Lactobacilli* include the need for lactose (Donaldson, 1964 and Hawley *et al.*, 1959). and sugar specificity. Pathogen colonisation may also be inhibited by certain carbohydrates, e.g. mannose (see explanation in Chapter 8) and lactose inhibit *S. typhimurium*.

Other natural compounds such as polypeptide lectins, are important in controlling bacterial adhesion as they cross-link the carbohydrate surfaces of bacteria with those of the target cell (see Chapter 8).

It has been shown that bacterial colonies grow in adherent glycocalyx-enclosed microcolonies (Cheng *et al.*, 1977). Cells are held in the nutrient dense area that allows them to collect and use soluble nutrients, and also gives them some protection from antibacterial compounds like chemicals, surfactants, antibodies and leucocytes.

Bacterial type in the gastro-intestinal tract

Prokaryotic cells can be classified according to microscopic shape (cocci, rods, spirilla), Gram staining and whether they are aerobic or anaerobic (Table 4.4).

Volatile fatty acids and their role in determining the gastro-intestinal flora

Volatile fatty acids (VFAs) are toxic to Gram-negative bacteria. Work by Meynell (1963) has shown that the gastro-intestinal micro-flora is controlled by VFAs. His work

Plate 4.1 Lactic acid bacteria adhered to a starch granule. [*By courtesy of* FMBRA.]

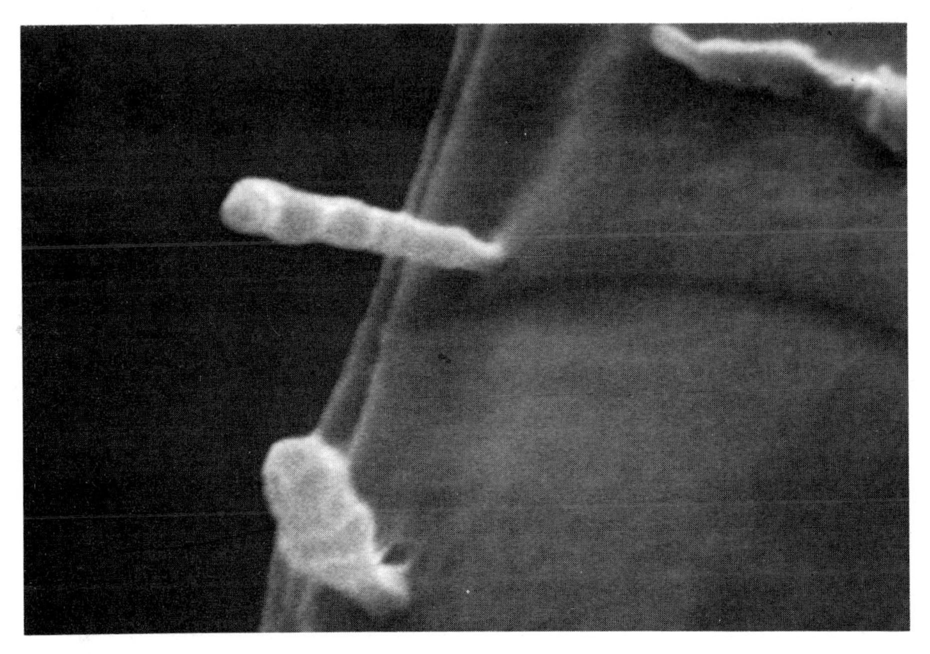

Plate 4.2 Point of adherence of lactic acid bacteria to a starch granule. [*By courtesy of FMBRA*].

Table 4.4 *Classification of prokaryotic cells.*

COCCI (spherical bacteria)

1. Gram positive cocci
   Aerobes – *Pediococcus, Staphylococcus, Streptococcus*
   Anaerobes – *Peptostreptococcus, Ruminococcus*
2. Gram negative cocci and coccobacilli
   Aerobes – *Acinetobacter*
3. Gram negative cocci
   Anaerobes – *Acidaminococcus, Megasphaera*

RODS (straight, cylindrical bacteria)

1. Gram positive, non-spore-forming rods
   Aerobes – *Lactobacillus, Listeria, Erysipelothrix*
2. Coryneform bacteria and actinomycetes
   Aerobes – *Corynebacterium, Arthrobacter, Brevibacterium, Cellulomonas, Eubacterium, Bifidobacterium, Mycobacterium, Actinomyces, Streptomyces, Streptosporangium*
3. Endospore-forming rods and cocci
   Aerobes – *Bacillus, Sporolactobacillus*
   Anaerobes – *Clostridium*
4. Gram-negative aerobic rods and cocci
   Aerobes – *Pseudomonas, Acetobacter, Agrobacterium, Brucelloa, Legionella*
5. Gram-negative, aerobic, chemolithotrophic bacteria
   Aerobes – *Nitrobacter, Nitrosomanas, Nitrosospira, Nitrosococcus, Thiobacillus, Thiobacterium, Thiovulum*
6. Sheathed bacteria
   Aerobes – *Streptothrix Crenothrix*
7. Gram-negative, facultatively anaerobic rods
   Facultative Anaerobes – *Escherichia, Klebsiella, Enterobacter, Salmonella, Shigella, Proteus*
8. Gram-negative anaerobic bacteria
   Strict Anaerobes – *Bacteroides, Fusobacterium*
9. Methanogens and other archaebacteria
   Methanobacterium, Methanococcus
   Aerobes – *Halobacterium, Halococcus*
   Anaerobes – *Thermoproteus*

CURVED RODS AND FLEXIBLE CELLS

1. Gram-negative spirillar and curved bacteria
   Aerobes – *Sprillum, Azospirillum, Campylobacter*
2. Gram-negative curved, anaerobic bacteria
   Anaerobes – *Succinivibrio, Selenomonas*
   Aerobes And Anaerobes – *Spirochaeta, Leptospira*

LARGE SPECIAL GROUPS

1. Gliding bacteria (always gram-negative) – *Myxococcus*
2. Bacteria with appendices and budding bacteria – *Hyphomonas*
3. Obligate parasitic bacteria – *Rickettisa*
4. Mycoplasma group – *Mycoplasma*
5. Anaerobic, anoxygenic phototrophic bacteria – *Rhodospirillum, Rhodopseudomonas, Rhodomicrobium*
6. Aerobic, oxygenic phototrophic bacteria – Cyanobacteria

(Based on Schlegel H.G., 1986)

demonstrated that the *in vitro* multiplication of *S. enteridis* was inhibited by VFA from the intestinal contents of conventional animals.

Morphology of unicellular bacteria

The gastro-intestinal tract is host to a wide variety of different shaped and arranged bacteria. Cocci are spherical cells and are found in irregular clusters, e.g. *Staphylococcus*, in regular arrangements, e.g. *Sarcina ventribuli*, or in chains, e.g. *Streptococcus pyrogenes*. Other shapes of cocci exist, e.g. diplococci, which are flattened (see *Figure 4.7*). The different arrangements of bacteria provide protection to the organisms in their specific habitats.

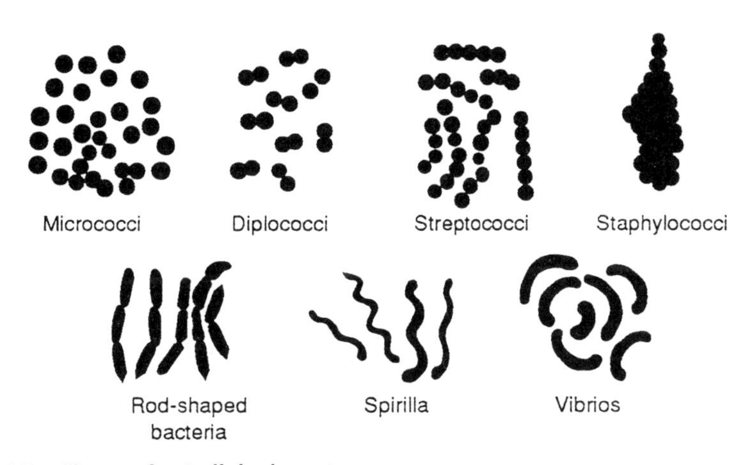

Figure 4.7 Shapes of unicellular bacteria.

The role of micro-organisms in the gut

Autochthonous bacterial population in the gastro intestinal tract

In addition to the bacteria that come into contact with animal tissues by the ingestion of food and water, and by the inhalation of air, many animal tissues attract and maintain specific bacterial species on their surfaces, and some even establish physiological symbiotic relationships with these truly autochthonous bacteria (Cheng *et al.*, 1981). Autochthonous bacteria have been defined as those that colonise particular tissues, usually in a succession from infancy to maturity of the host, and maintain stable population levels of normal flora of healthy adults (Savage, 1977). The type of growth of autochthonous bacterial populations is important because these organisms persist for long periods of time in immediate contact with the surface, and they must therefore be capable of withstanding the operation of a continuously stimulated host defence system, against whose specific agents (i.e. surfactants, antibodies, phagocytes) they have no special protection.

Changes in the bacterial colonisation of digestive systems
Some animal digestive systems are found to be colonised by a simple species of bacteria (Savage, 1977) . However, monospecies colonisations tend to be on tissue surfaces in the stomach where conditions already limit the number of bacterial species that can survive and grow.

In most areas of the digestive tract there are many different species of bacteria. The adherent autochthonous bacterial population makes an important physiological contribution to the health of the whole animal. Data suggest that the 400 μm "unstirred" layer on the surface of intestinal tissues (Levin, 1979) comprises a biofilm of bacteria, protozoa, mucus and exopolysaccharides and that this very thick layer constitutes a complex viscous ecosystem within which a potential pathogen must gain a measure of dominance before it can approach and colonise the tissue surface.

As described elsewhere different bacterial adhesion mechanisms also exist, and even the pilus-mediated adhesion of *E. coli* in the neonatal ileum is complicated because strains lacking all of the well-recognised pili can cause fatal diarrhoea in colostrum-deprived calves (Chan *et al.*, 1982b). Other difficulties include the presence of mucus and serum factors on the surface of the target tissue, with many autochthonous bacteria actually binding to fibronectin on the cell surface (Woods *et al.*, 1980). However, many pathogenic bacteria live in the mucus at the cell surface and react only secondarily with the target cell.

The proteins of the bacterial cell surface may be located partially within the outer membrane, in an extracellular protein coat, or in protruding fimbriae. They protrude through the glycocalyx and are clearly available to bind to target cell ligands at a considerable distance (1 to 6 μm) from the bacterial surface.

Indigenous bacteria associated with epithelial surfaces are important in exerting certain influences by which the host animal resists some microbial diseases. Such organisms have long been known to inhibit non-indigenous bacteria of related and unrelated species from colonising niches in epithelial habitats. This phenomenon, known as "bacterial interference", has been studied intensively by numerous investigators in recent years (Barrow *et al.*, 1980; Freter, 1983). Evidence now supports a hypothesis that the interference phenomenon has many possible mechanisms and may be exerted by the indigenous micro-flora either directly or indirectly.

In modern day intensive agriculture there are a number of factors which affect the introduction and development of the normal flora. The most important of these are: excessive hygiene, antibiotic therapy and stress (Fuller, 1989).

Stress

The disturbance of this natural, bacterial, barrier function by manipulation, trauma, catheterisation, broad-spectrum antibiotic therapy, or hormonal changes may allow an ascending infection by faecal pathogens such as *E. coli*. Stresses such as underlying disease and trauma (e.g. surgery) alter the protease content of saliva, and Woods *et al.* (1980) have shown that the fibronectin and the autochthonous bacterial population are lost from the oropharyngeal surface in stressed individuals. This autochthonous population is rapidly replaced by a biofilm composed largely of cells of *Pseudomonas aeruginosa*. Experiments with tracheal rings have shown that these organisms grow in a glycocalyx-enclosed biofilm on the colonised tissues (Baker and Marcus, 1982).

The influence of different forms of stress on intestinal micro-organisms is given in Table 4.5. The effect of stress on the gut micro-flora has been discussed by Tannock and Savage (1974), the general trend being for *Lactobacilli* to decrease and coliforms increase. Stress can be caused by any major change in the physical or emotional environment. It can cause secretion by the pituitary gland of adrenocorticotrophic hormones which stimulate the adrenal cortex to synthesise corticoids.

Several reports (Tannock and Savage, 1974; Kenworthy and Crabb, 1963; Schulman, 1973) have observed decreased *Lactobacilli* and increased coliforms when animals have been subjected to a stress such as weaning, diet change and transportation.

Weaning

It has been shown that weaning, for example of pigs, can cause changes in gastric function that accelerate the growth of *E. coli*, a normal bacterial inhabitant of the digestive tract that increases in number during diarrohea (Schulman, 1973). Several researchers (Muralidhara *et al.*, 1977; Mitchell and Kenworthy, 1976; Porter and Kenworthy, 1969; Moon, 1975) have shown that *Lactobacilli* reduce pathogenic gut coliforms, which may be an important part of post-weaning growth check commonly seen in young pigs.

The balance of pathogens and non-pathogens

In recent years, interest has grown in the feeding of *Lactobacilli* due to the high use of antibiotics. Antibiotic therapy often lowers the *Lactobacillus* population in the intestinal

Table 4.5 *Some reported examples of the influence of stress on intestinal micro-flora.*

Stress	*Species*	*Effect*	*Author*
Starvation	Rat	Decreased *Lactobacilli*	Smith (1965)
Antibiotic therapy	Human	Decreased *Lactobacilli* Increased *Staphylococcus*	Gordon (1955)
Starvation	Poultry	Decreased *Lactobacilli*	Fuller and Turvey (1971)
Old cages	Poultry	Increased coliforms	Tortuero (1973)
Starvation	Mouse	Increased coliforms	Tannock and Savage (1974)
Loss of food/water	Mice	Increased in *Lactobacilli* Decreased in coliforms	Tannock and Savage (1974)
Transport		Reduced *Lactobacillus*	Hutcheson *et al.* (1980)
Transport		Increased coliforms	McCormick (1984)

tract and it is claimed by most probiotic producers that replenishing the intestinal tract with *L. acidophilus* results in an accelerated return to normal intestinal micro-flora. In a healthy animal a proper balance of the various normal inhabitants of the intestinal tract is needed. However, it is claimed that modern day farming often stops the young animal from obtaining its normal bacterial loading. It has been suggested that the most satisfactory method to overcome this is to feed bacteria. Best results are claimed to be obtained from the ingestion of 1×10^8 to 1×10^9 viable *L. acidophilus* daily; ingestion of excessive numbers may induce mild diarrhoea.

A relationship between *Lactobacilli* (L) and coliforms (C) has been proposed by expressing the data as a *Lactobacilli* to coliform ratio (L:C) (Chopra *et al.*, 1963; Muralidhara *et al.*, 1977). It is suggested that the higher the L:C ratio is, the better the micro-floral contribution to growth performance of host.

While the precise method(s) by which *Lactobacilli* and other desirable micro-organisms produce beneficial effects within the gastro-intestinal tract of animals and regulate other components of the intestinal flora is uncertain, it should be remembered that they are themselves the microbial group most likely to be influenced adversely by the stresses of an altered environment (Fuller *et al.*, 1977). The number of *Lactobacilli* and other beneficial types of bacteria decreases in times of stress and consequently, so will their beneficial effects, allowing pathogens such as coliforms to increase (Muralidhara *et al.*, 1973; Chopra *et al.*, 1963; Hill *et al.*, 1970). It should be noted that the levels of *Lactobacilli* normally resident in the alimentary tract increase as an animal gets older (Muralidhara *et al.*, 1973). Any product designed to alter their numbers is less likely to show an effect in mature animals because they are more likely to have a well established balance of micro-organisms and are less likely to suffer from minor digestive disturbances. Taking the newborn piglet as an example, minimising pathogen levels in the gut can be attained by:

- Lowering the level of potential pathogens in the piglet's immediate environment by disinfection, etc.
- Passive protection by immunisation of the sow with an *E. coli* vaccine prior to farrowing.
- Administration of direct fed microbials to the piglets themselves immediately after birth to encourage gastro-intestinal colonisation by lactic acid bacteria for their protective effects against pathogenic coliforms.
- The administration of organic acids, such as lactic acid, would make the gastro-intestinal conditions of the piglet less suitable for coliform proliferation and so may also benefit the piglet in terms of its gastro-intesinal microbial balance.

The immune system

The mechanisms of the body that react passively or actively against foreign materials constitute the immune system. It represents a defence system to attack by harmful materials. Many factors contribute to the immune system in some form of protection against invading organisms, for example, immunisation by passive serum or actively by vaccine. Perhaps two of the most significant milestones were; first, work by Jenner on vaccination against smallpox in 1796 by innoculation with cowpox; second, the work of Pasteur and his colleagues who developed a range of vaccines and directed work towards the study of resistance through the blood, and the production of antibodies. Work at the Pasteur Institute by Metchnikoff and his colleagues showed that resistance to infection was not solely through antibodies of the blood. They showed in the rabbit, that resistance to infection in the gut was more closely related to local antibodies than to antibodies in the blood.

Immune response in the gut

The gut faces many and varied challenges from antigens (e.g. bacteria, viruses, parasites and a number of nutrients) and is able to respond with the mucosal defence system secreting immunoglobins (antibodies). The immunoglobulins (Ig) are produced by lymphocytes and are the major factor in the defensive response in the gut. They are a family of proteins having a similar structure. They are made up of light and heavy chains linked by disulphide bonds to give a Y structure. It is the shape at the end of the arms that varies from one molecule (the variable area) to another to give a wide range of antigen building sites.

Members are divided into classes (e.g. IgA, IgE) according to their heavy chains. Some of particular importance are IgM, IgA and IgG. IgA is the major local secretory immunoglobin. It is produced by cells in the intestinal tissues and represents the first line of defence. IgG is blood-borne and is the second line of defence. It is produced in such organs as the lymph nodes and spleen and gets into the gut when it is damaged.

The young animal

In some species, e.g. the human, antibodies are able to cross the placenta and allow the young to be born with some degree of protection. However, this is not the case with all species (e.g. the pig) and they are born with no circulating antibodies. In this case the milk of the first few days (colostrum) is very high in immunoglobins (IgA). Although the level falls, it is produced throughout lactation. The immunoglobins are large molecules and the newborn piglet has the ability to absorb them in an intact form for a short period after birth (*Figure 4.8*). The IgA is well suited to the young animal as it is fairly resistant to the digestive enzymes and can also attach to the gut to give a defence against pathogenic challenge. By about the third week in life the piglet is able to produce its own secretory IgA.

"Oral immunization" has been used to give protection against microbial pathogens. For example, enteric diseases associated with *E. coli* have been controlled by in-feed treatment of the pre-partum sow and post-weaning piglet (Porter *et al.*, 1985). Similar techniques have been used in the control of *Salmonella* in the calf (Balger *et al.*, 1981) and coccidiosis in poultry (Davis *et al.*, 1986).

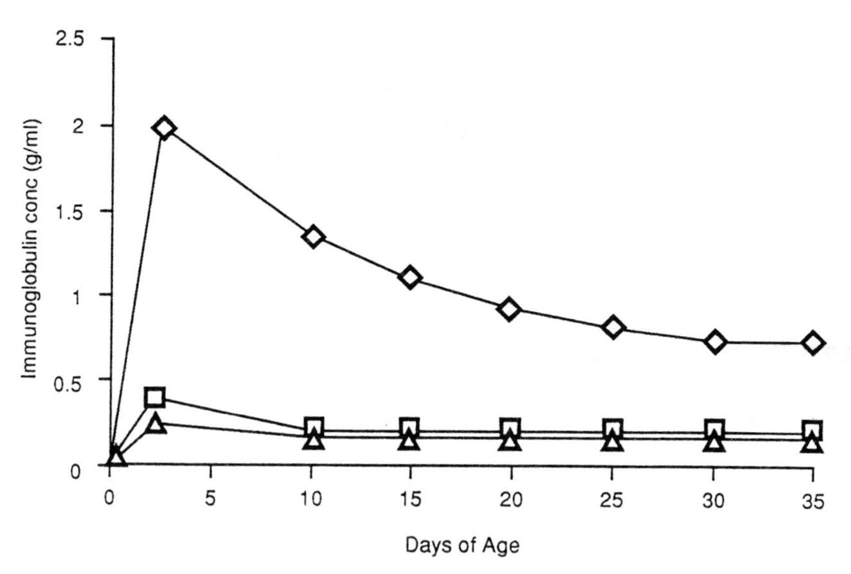

Figure 4.8 Serum immunoglobulin levels in colostrum fed pigs (◇ IgG □ IgA △ IgM) (after Porter and Hill, 1970).

The gastro-intestinal tract and immunity

Mucosal non-immunoglobulin immunity

Mucosal surfaces (lungs, gut and the respiratory, reproductive and urinogenital tracts), incorporate an antimicrobial enzyme system as a primary defence against microbial attachment. It is effectively independent of, and precedes, the activation of the blood borne immunoglobulin response. Activity is non-specific and its purpose is to prevent the adhesion and subsequent invasion of opportunist bacteria, viruses, and fungi (Jacobs *et al.*, 1972; Reiter, 1979; Pruit and Reiter, 1985). Effective functioning of this mucosal antimicrobial barrier is dependent on a supply of hydrogen peroxide, appropriate halide ions and peroxidases. Hydrogen peroxide can be a limiting factor. Halide ions are usually freely available but are diet dependent (Reiter, 1985) and peroxidase production varies according to the time in the sexual cycle, summer or winter season, diet and breed, etc. (Kern, Wildbrett and Kiermeier, 1962).

Stress and immunity

Stress can be environmental or psychological and is known to depress the immune response in both animals and humans. It has been demonstrated that heat stressed sows have elevated cortisol levels, and IgG levels in sow's serum decrease at parturition. Higher cortisol concentration in serum and lower IgG in colostrum of sows under heat stress was associated, in their piglets, with higher serum cortisol at birth and lower serum IgG for the first 20 days post partum (Machado-Neto, *et al.*, 1987). Depressed immune response has also been demonstrated in piglets weaned younger than five weeks of age (Blecha *et al.*, 1983), and in transported cattle (Blecha *et al.*, 1984).

Antibiotics and immunity

The effect of antibiotics on the digestive micro-flora depends on the type and level of the antibiotic used. Some have been shown to depress the immune response significantly, at the same time resulting in a significant reduction in the weights of the spleen and thymus. In such cases antibiotic withdrawal results in restoration of the gut flora and in a return of the immune function.

Germ-free or gnotobiotic animals

In normal commercial and natural circumstances young animals are colonised by bacteria very shortly after birth. The first colonisers are generally the haemolytic clostridia with other organisms following and achieving highest levels in the caeca and crop of chickens.

Germ-free animals are obtained by various techniques, e.g. hatching, hysterectomy and hysterotomy followed by rearing in an environment totally free of detectable micro-organisms, leaving a practically 'sterile' gastro-intestinal tract. Germ-free animals have been associated with the ultimate case of antibiosis. However, today antibiotics are usually selective against a particular species of bacteria and they may not produce a general reduction in bacterial loading.

Germ-free chickens have been shown to grow better than conventional chicks, and even with the administration of oral antibiotics to normal animals, performance was not improved to an equal level (Forbes and Park, 1959; Coates *et al.*, 1963). However, administration of fresh excreta from conventional poultry, reduced the growth rate of germ-free birds. This improvement in growth of germ-free chicks over conventional animals indicates that the beneficial effects of the total volume of micro-organisms are outweighed by the detrimental effects (Coates, 1976; Ratcliffe, 1985).

Other characteristics of germ-free animals include:

- Less K, Na, Cl in caeca (Asano, 1967).
- Higher level of alkaline phosphatase activity (Jervis and Biggers, 1964).
- Smaller mucosal cell turnover (Visek, 1978).
- Longer villi in duodenum (Abrams *et al.*, 1963).
- Increased nitrogen loss from the gut, e.g. urea, amino-acids, peptides, muco-proteins, which would normally be metabolised (Salter, 1973).
- Lower levels of IgA in the serum of un-colonised animals (March, 1979).
- Reduced lymphoid follicles and free plasma cells (Gordon, 1955; Thorbecke *et al.*, 1954).
- Thinner intestinal epithelium associated connective and lymphoid tissue.
- Different absorptive surface area of the small intestine (Rupin *et al.* 1980; Savage, 1985; Savage and Whitt, 1982; Whitt and Savage, 1981, 1988) due to longer villi, etc.
- Production of D-leucine which has inhibitory properties (Gilliland and Speck, 1968).

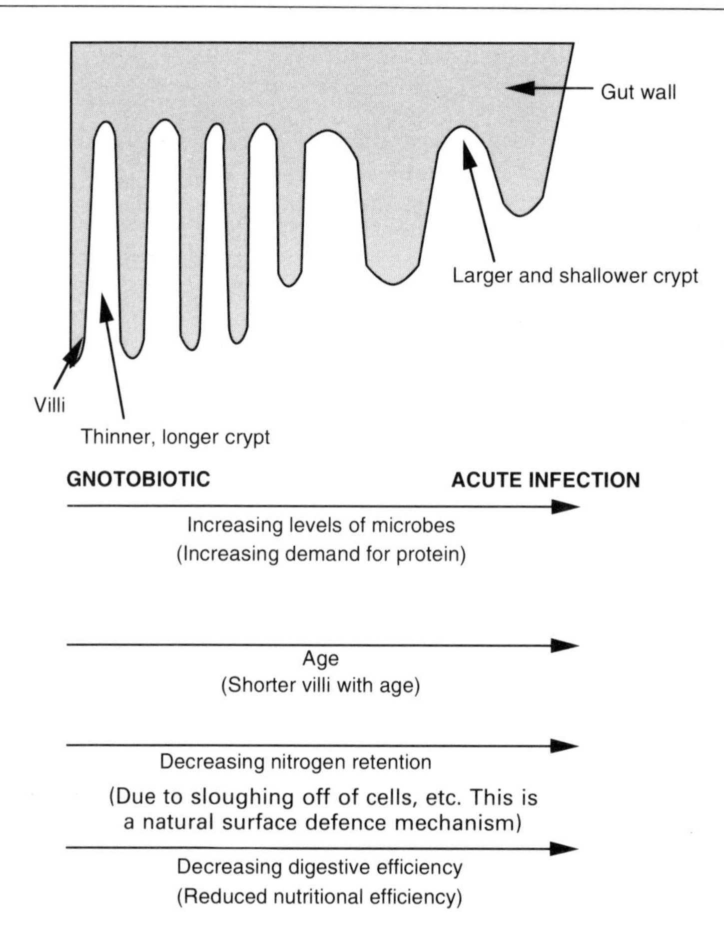

Figure 4.9 Illustration showing the change in villi length, with age and microbial loading.

- Production of lactate (Young *et al.*, 1956; Alexander and Davis, 1963).
- Longer transit time of ingesta.
- Enlarged caecum, accumulation of muco-polysaccharide material, a loss of muscle tone and an influx of water to the caecum (Wostmann *et al.*, 1973).
- Better utilisation of energy (Furuse and Yokota, 1985).

Germ-free animals have a more regular and thinner villus structure with narrower lamina propria. (see *Figure 4.9*). The total weight of the small intestine is lower in germ-free animals (Coates, 1980), with a slower renewal rate of epithelium, a lower daily metabolic energy cost and a consequent decrease in endogenous losses. When bacteria are present inflammation and thickening occurs and reduces the efficiency and absorptive capacity of the small intestine (Sprinz, 1962).

Effect of the metabolic activity of gut micro-organisms

Bacterial fermentation in the gut may result in the reduction of nutrients available to the host due to active metabolism of ingested nutrients and loss of bacterial protein, etc. in the faeces. It is therefore important, if the bacteria present are using nutrients, that, at least, they are beneficial in other terms to the host. A summary of the benefits and negative effects of the gut micro-organisms is given in Table 4.6.

Table 4.6 *The effects of gut micro-organisms.*

Benefits	*Negative Effects*
Synthesis of vitamins B and K	Production of toxic metabolites
Detoxification of food components or endogenous products	Modification of nutrients
Recovery of endogenous nitrogen	Release of toxins from non-toxic precursors
Production of digestive enzymes, e.g. bacterial amylase for starch digestion	Uptake of nutrients, e.g. amino-acids Decreased digestibility of fat due to altering lipids and bile salts

(Coates, 1980)

The micro-flora of the gut can be split into two components, those occurring free or attached to the food particles in the lumen and those associated with epithelium (Dubos *et al.*, 1965). Those associated with epithelium are likely to have more effect on the thickening and increase of mass of the intestinal epithelium in normal animals.

Gut micro-flora can cut down fat absorption by enzymatically removing glycine or taurine from conjugated bile salts. This results in lowering the concentration of conjugated salts below the critical level required for micelles to form.

Metabolism takes place in avian caeca as a result of the micro-flora but there is little evidence to support caecal absorption of bacterial metabolites (Barnes and Impey, 1974). A large number of these anaerobic bacteria are found in the caeca (up to 10^{10}/g of contents) and are able to decompose uric acid (Barnes *et al.*, 1972; Mead and Adams, 1975).

CHAPTER 5

Infections of the digestive tract

Introduction

Various forms of enteritis and diarrhoea are common in animals and man. For example, after weaning, a number of farm livestock exhibit diarrhoea which is associated with production of enterotoxins by *E. coli* and *Clostridium Spp*. The move to a dry food gives a 'carbohydrate' overload where undigested carbohydrates provide a medium for these potentially harmful bacteria. At the same time, the lactic acid producing bacteria are adversely affected by the removal of the milk diet.

In this book two particular organisms, *E. coli* and *Salmonella*, have been chosen for detailed consideration because of their widespread involvement in problems in the digestive tract. Not only do they cause problems to humans but they present major problems in the livestock industry.

E. coli and diarrhoea

E. coli is a non spore-forming, medium-size bacillus. It is a cylindrical bacterium about 2–3 μm (1/5000th of an inch) long and 0.5 μm wide with rounded ends. Some species are capsulate and the majority are motile. The very thin projections of *E. coli* (called pili or fimbriae) act as anchors to the gut surface receptors (*Figures* 5.1 and 5.2). They comprise lectins on the surface of the bacteria which recognise carbohydrate residues. These fimbriae form the basis of the antigenic structure of *E. coli*. The various serotypes may be identified very precisely depending on the O, K or H antigens. *E. coli* with capsules covering the outer membrane of the typical O antigen are K antigens (K is derived from the German kapsel). They are numbered, K1, K2 . . . etc. and there are at least 89 K antigens. Their adherence property makes them very virulent and

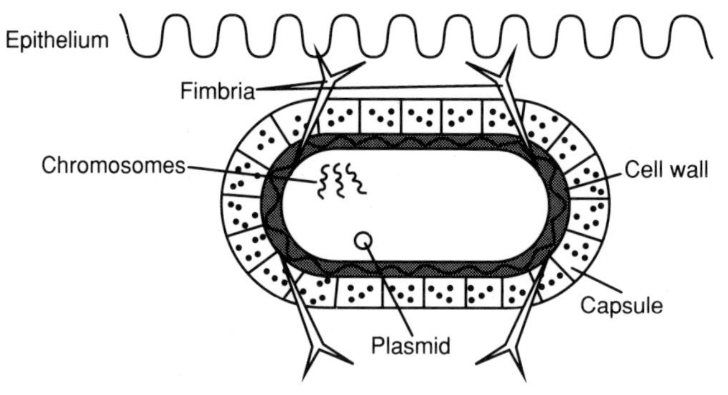

Figure 5.1 *E. coli* attachment to epithelial cells by use of fimbriae.

they include K88 and K99 which are associated with diarrhoea in young animals. H antigens are flagellar but are poorly developed in *Escherischia*.

E. coli are found everywhere and even attach to the dust particles in the air. Relative to particle size, greater numbers of *E. coli* attach to the smaller-sized particles than larger ones. These are the most numerous in the air, stay suspended the longest, and are most likely to penetrate the deeper parts of the respiratory tract (lungs and airsacs) when inhaled by the birds and animals (Barnes, 1987). In poultry, the main site for *E. coli* and *Salmonella* colonisation are the crop and caecum (Weinack *et al.*, 1981). The normal gastro-intestinal tract of domestic animals maintains a stable flora (Donaldson, 1964) and it is only when the gut is disturbed or a very pathogenic *E. coli* strain is present that the balance is tipped.

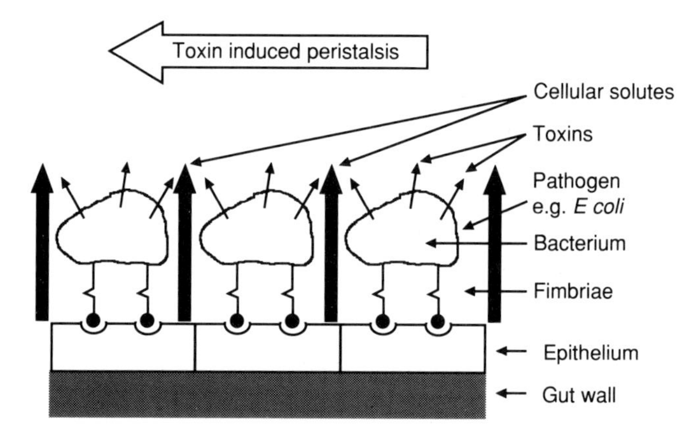

Figure 5.2 Diarrhoea in young animals is often caused after *E. coli* overgrowth. It is only when pathogens attach in the gut wall that they produce the toxins associated with diarrhoea. The toxins result in an increased rate of peristalsis.

Bacterial translocation

The body naturally possesses defence mechanisms to prevent indigenous pathogenic bacteria from crossing the gut wall into the body. The mechanisms are essential as pathogenic organisms are more likely to have lethal effects when they have crossed the gut wall. These defence mechanisms have been reported as:

- A complete intact mucosal layer to act as a barrier to entry.
- A healthy population of non-pathogenic gastro-intestinal micro-flora to prevent overgrowth by pathogenic bacteria (Steffen and Berg, 1983).
- A natural host immune defence system.

Further information on some of these factors is given in Chapters 4 and 7.

Pigs

Colibacillosis (baby pig 'scouring' or diarrhoea caused by *E. coli*) is a major source of economic loss to the livestock industry. Losses result from both increased mortality and reduced growth rates. It has been found that acutely ill pigs have low *E. coli* levels in the stomach but high levels in the anterior portion of the small intestine. Enterotoxigenic *E. coli* strains cause diarrhoea by attaching to the intestinal mucosa and by producing enterotoxins that cause an influx of sodium ions and water into the intestine.

Post-weaning diarrhoea in pigs has been associated with increased metabolic activity of *E. coli* in converting proteins to amines. Since amines are irritating and toxic they increase intestinal peristalsis and produce diarrhoea (Porter and Kenworthy, 1969) (*Figure* 5.2). It should, however, be noted that the pathogenicity of *E. coli* is restricted to certain strains, with the vast majority of strains causing no ill effects. Colibacillosis is caused by the invasion of the anterior small intestine of the piglet by pathogenic strains of *E. coli* that possess at least two virulence factors. Organisms possess colonisation antigens or fimbriae (adhesins or pili) that enable the bacterial cells to recognise and attach to specific mannose receptor sites on the mucosal brush border (*Figure 5.1*). In addition, bacteria must produce enterotoxins which act upon the secretory cells located in the crypts of the villi. *S. typhimurium* and *E. coli* have been suggested to share common receptor sites and a common mechanism for attachment (Soerjadi *et al.*, 1981) .

The attachment of the pathogenic *E. coli* to the mucosal cells enables them to overcome the mechanical clearance of the intestine caused by peristalsis, and facilitates colonisation of the gastro-intestinal tract by these harmful organisms. Once the bacteria have colonised the gut, they start to produce the enterotoxins. There are two main classes of toxin:

- *Heat labile toxin* characterised by a high molecular weight
- *Heat stable toxin* characterised by a lower molecular weight

The toxins inhibit the absorptive processes within the intestine, thereby increasing liquid levels in the lumen. The villi become much shorter and consequently have a smaller surface area for absorption. The volumes of fluid are too great to be re-absorbed from the large intestine and this consequently causes ionic imbalances and the development of diarrhoea (*Figure* 5.3).

The piglet is born free from circulating antibodies and obtains them with the first feeding of colostrum (Underdahl, 1983). Sow vaccination with either live *E. coli* or a typical antigen e.g. K88 has been found to successfully reduce neonatal piglet mortality (Chidlow, 1979). Pigs pretreated with a K88+ non-pathogenic strain of *E. coli* were more resistant to infection with a pathogenic strain of *E. coli* which shared the K88 adhesin. The treated group suffered less diarrhoea and fewer animals died.

Villi in the intestine of the piglet increase in length up to 10 days of age, but appear to decrease after this. This reduction may be due to the nature of the feed being more abrasive and concentrated. These, generally, solid feeds are taken in large quantities after weaning. However, a whole series of events is associated with the change of nutrition at this stage. The importance of dietary lactose has been stressed by Powles and Cole (1993) who also illustrated its role as a substrate for *Lactobacilli*. *E. coli* are normal gut inhabitants whose numbers increase immediately after weaning, with the risk of a pathogenic strain developing (McAllister *et al*, 1979). When a pathogenic strain establishes, toxin production starts and fluid secretion from the intestinal wall increases. This is followed by depletion of body electrolytes, dehydration, acidosis and finally death.

Pathogenic bacteria, however, directly reduce cell effectiveness and lead the animal to increase the villi number to compensate for this loss in villi length. In the small intestine they cause damage either by producing toxins or by entering the enterocyte.

The piglet, for example, is more susceptible to colibacillosis than the fowl, due to a lower internal body temperature and reduced gut mobility which encourages luminal adherence (Arp and Jensen, 1980). Antibiotics can be used to treat colibacillosis but they may accentuate the problem by facilitating the aberration of endotoxin due to cell disintegration.

Poultry

E. coli infection is an example of a "production disease" which has become increasingly important as production of poultry has intensified. A considerable amount of money is lost annually as a result of *E. coli*. For example, a 1985 survey of turkey health in Minnesota showed that over 40% of the loss from disease could be directly or indirectly attributed to *E. coli* infection (Barnes, 1987). In the USA alone, 4% of all broiler chickens and 6% of all turkeys die from *E. coli* related disease every year (Barnes, 1987).

In poultry, the crop and caecum are the main sites of colonization for *E. coli* and *Salmonella*. The presence of non-pathogenic bacteria can be associated with a lengthening of the villi and increased turnover of their epithelial layers (Cook and Bird, 1973). Deeper crypts appear, and there is an increase in the rate of cell migration. This change in the villi of chickens was found to be more prominent in the upper than lower parts of the small intestine. The native gut micro-flora can reduce colonisation by *E. coli*

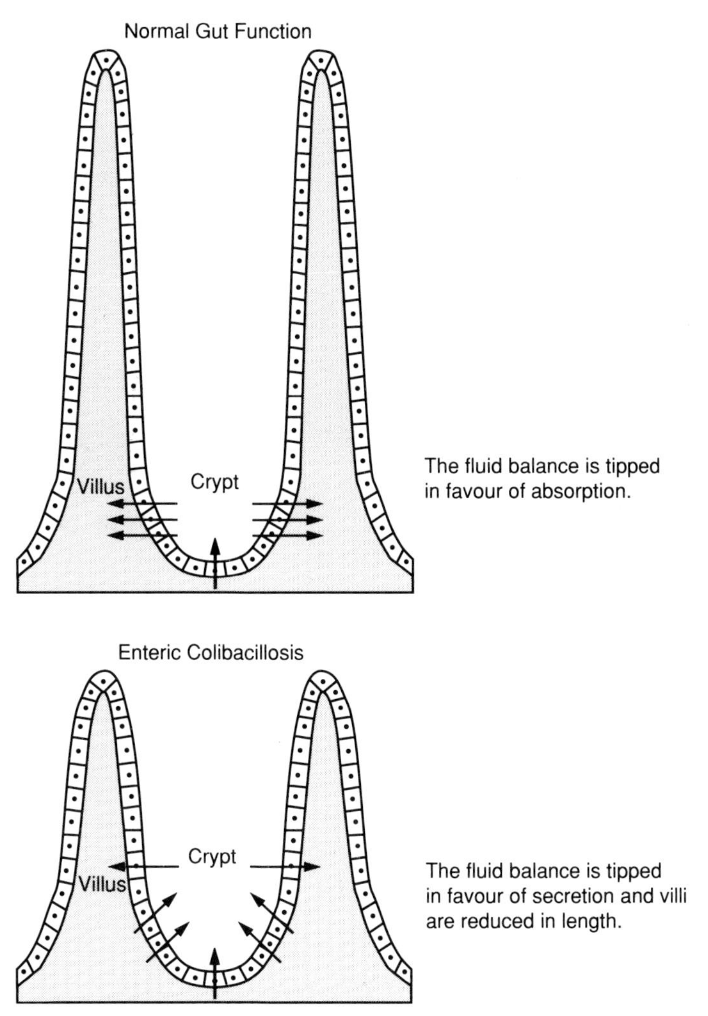

Figure 5.3 The villi of healthy pigs and those which have become shortened during colibacillosis.

at these sites (Weinack *et al.*, 1981) by competitive exclusion and this can be enhanced by probiosis.

Two other important effects of *E. coli* have been reported in turkeys in two different ways. Firstly, they may infect the respiratory system. A cheese-like residue forms over the air sacs and induces death due to lack of oxygen. Birds are susceptible from 3 weeks of age until slaughter. Secondly, *E. coli* septicaemia, usually considered a secondary disease, may be precipitated by another disease or stress. The systemic nature of this disease includes symptoms such as an enlarged spleen and liver damage. Birds die very quickly with *E. coli* septicaemia, while with *E. coli* infection of the respiratory tract, the effects last for several days. *E. coli* septicaemia usually occurs between 42 and 84 days of age.

Calves

Diarrhoea is a common occurrence in calves. Probably much of it is not infectious in origin but is the result of the artificial manner in which many calves are reared. This type of diarrhoea features prominently among sick animals. High *E. coli* levels are often found in the abomasum of animals suffering from diarrhoea (Ingram, 1962).

For a strain to be able to produce diarrhoea in a particular species of animal, it must possess at least two properties; the ability to produce an enterotoxin active in the small intestine, and an ability to proliferate in the small intestine of that species.

Neonatal diarrhoea in calves has been reported to be due to colonisation of the intestine by *E. coli* strains producing K99 and/or F41 fimbriae (Gaastra and DeGraaf, 1982). More recently however, other fimbriae have been reported in Belgium (Pohl *et al.*, 1982; 1984; France, Contrepois and Girardeau, 1985; Morris *et al.*, 1985). Epidemiological studies have revealed that *E. coli* F17+ strains are associated with other outbreaks of coligenic diarrhoea such as enterocolitis. In summary, therefore, pathogenic cells have adapted different fimbriae to colonise to different receptors on the gut epithelium (see *Figure 5.4*).

Salmonellae

Salmonella has come under increasing public scrutiny as a food poisoning organism over the past few years. Clinical illness in humans caused by the *Salmonella* organism ranges from a mild stomach upset, to headaches, to nausea and abdominal pain, to-severe illness requiring medical attention over a prolonged period.

In 1990, 68 people died from *Salmonella* related diseases in England and Wales alone. Today over 2,200 serotypes of *Salmonella* that cause the upset in man are known. In the same report it was claimed that the number of cases of *Salmonella* isolated from human faeces had doubled from 1981 to 1991 (10,251 to 20,532 in England and Wales). The most commonly isolated strains in the UK were *Salmonella enteritidis*, *S. typhimurium* and *S. virchow*. With two exceptions, all serotypes are motile. They have a morphology and cultural requirement very similar to those of *E. coli*.

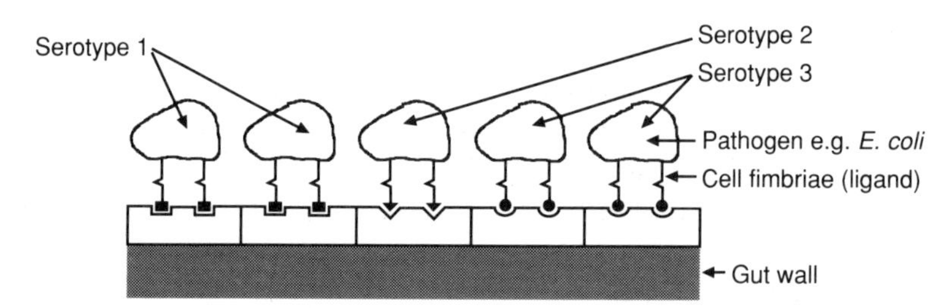

Figure 5.4 *E. coli* attachment to different epithelial cells following the development of specific fimbria.

Table 5.1 *The level of* Salmonella *contamination in common feedstuffs as a % of screened samples.*

Fish meal	2.9
Meat and bone	2.7
Extruded soya	2.3
Broiler feed	0
Layer feed	0

(Watson, 1989)

Salmonellae enteritidis now accounts for more than 36% of the reported *Salmonella* infections (Anon, 1993). The organism is rarely found in ruminant animals with most strains occurring in poultry (phage types 4,6,8). *Salmonella enteritidis* can be found in high numbers in the caeca (Brownell *et al*, 1970), even when the animal may show no clinical signs of infection, nor suffer any performance effect.

Although animal feedstuffs are a main source of infection, it has been shown that the level of contamination is very low (Watson, 1989). (See Table 5.1). Agricultural feedstuffs, and eggs, in particular, have been highlighted as the main source of the contamination in humans with little evidence to support it. However, other materials have from time to time been affected on a large scale, for example, tomatoes in the USA. It is just as likely that the infections may arise from poor storage, processing and cooking of foodstuffs.

The levels shown in Table 5.1 have particular significance in relation to two orders introduced in the United Kingdom to control contamination of animal feed. They are "The Disease of Animals (Protein Processing) Order, 1981" and the "Importation of Processed Animal Protein Order, 1981". Contamination is commonly controlled by the use of organic acids.

Although levels of *Salmonella* contamination in feed are reported to be low, young chicks can become infected from feed containing only one *Salmonella* per gram, so every attempt must be made to eliminate these bacteria. The day-old chick is readily infected by *Salmonella* but develops a substantial resistance within the first few weeks of life, so that considerably greater numbers of *Salmonellae* are necessary to establish infection after this time. Modern poultry husbandry often precludes the development of proper intestinal micro-flora which may play a role in protecting birds against other enteric infections (Lloyd *et al.*, 1977; Snoeyenbos *et al.*, 1978).

Resistance of young chicks and poults to *Salmonella* infection can be substantially increased by early oral administration of intestinal contents of faeces from selected adult chickens (Rantala and Nurmi, 1973; Lloyd *et al.*, 1977; Snoeynobos *et al.*, 1979; Weinack, 1981; Soerjadi *et al.*,1981). These authors reported that the protective mechanism appeared to be a consequence of competitive exclusion of *Salmonella* by "normal" micro-flora of the gastro-intestinal tract. The higher susceptibility of young chicks, compared with older birds, to *E. coli* infections, was thought to be due to an immature protective micro-flora. "Normal" intestinal micro-flora created quite a stable ecosystem helping to inhibit the establishment of opportunistic pathogens such as *Salmonella*. Although there can be no doubt that the proper intestinal micro-flora of the chicken and turkey plays a significant role in protection against infection by at least some *Salmonellae*, the value of direct fed microbials to help prevent and control salmonella in these species is still unclear.

Contamination of birds from the environment has been reported to be the biggest source of infection by *Salmonella enteritidis* in poultry (Watson, 1979). Environmental sources including drinking water and even the slaughter house have been implicated (Patric *et al.*, 1973). A further source of contamination has been reportedly due to vertical transmission of *Salmonella enteritidis* where infected breeding stock have produced contaminated eggs which spreads the contamination to the grower flocks (Borland, 1975).

It is therefore important that the control of *Salmonella* in laying hens starts at the breeder and multiplier flock level. Elevated caecal *Salmonella* levels of up to four times normal values have been found in problem breeder flocks. Various management programmes have been suggested and include:

- Proper control and storage of floor shavings prior to use.
- The use of feeds that are pelleted at 90° for 10 seconds.
- Dedicated feed storage and haulage vehicles.
- Organic acids.
- Chlorination of drinking water.

CHAPTER 6

The use of antibiotics

The history of antibiotics

Antibiotics are chemical compounds, which in small quantities are harmful to other organisms. They occur widely in nature but the term is generally restricted to compounds which are microbial in origin, particularly those used to control other harmful micro-organisms (generally bacteria).

The first antibiotic to be described was penicillin, by Sir Alexander Fleming. It is a classic example of a major scientific discovery by serendipity rather than as the result of objective research. In 1928, a culture plate of *Staphylococcus* was contaminated with spores of *Penicillium notatum*. Around the mould which was contaminating the plate, the *Staphylococcus* was destroyed. The pure culture was isolated and found to produce a substance which had a powerful anti-bacterial effect, and was named penicillin. Later, Sir Howard Florey's team, at Oxford University, made particular studies of antibiotics in terms of their curative properties, isolation and production.

The last twenty years has seen a considerable intensification of the livestock industry, with faster growth rates, greater stocking density, larger production units, etc. The increased pressure on production has resulted in an increase in the use of low level feed additives to improve performance and/or health. Also, in intensive systems, the health of animals has been a problem, with high levels of stress and microbial challenge. Today, antibiotics are commonly used as feed growth promoters, as well as being used at higher levels for therapeutic purposes. Some of the more commonly used antibiotics are given in Table 6.1. However, in many countries there is critical examination of their use. Antibiotics have an essential role in agriculture; however, their indiscriminate use has even exacerbated some of the conditions that their development was intended to prevent.

The Swann Report (1969) recommended tighter governmental control on the use of antibiotics in feed in Britain and stated that antibiotics which produced cross-resistance

Table 6.1 *Some commonly administered oral antibiotics.*

Type	*Examples*	*Source*	*Mostly active against*
Penicillins and cephalosporins	Penicillin G	*P. chrysogen*	G+ bacteria
	Ampicillin	Semisynthetic	G+ and some G– bacteria
	Cephalexin	Semisynthetic	G+ and some G– bacteria
Tetracyclines	Chlortetracycline	*Streptomyces aureofaciens*	G+ and G– bacteria
	Oxytetracycline	*Streptomyces rimosus*	rickettsiae
	Tetracycline	*Streptomyces spp* or from chlortetracycline	G+ and G– bacteria rickettsiae
Macrolides	Erythromycin	*Streptomyces erythreus*	G+ bacteria
Aminoglycosides	Streptomycin	*Streptomyces grisens*	G+ and G– bacteria tubercle bacillus
	Neomycin B	*Streptomyces fradiae*	G+ and G– bacteria
Polypeptides	Bacitracin	*Bacillus licheniformis*	G+ bacteria
Others	Chloramphenicol	*Streptomyces venezuelae* or synthetic	G+ and G– bacteria including some *Salmonella rickettsiae*
	Lincomycin	*Streptomyces lincolnensis*	G+ bacteria

should not be used in animal feeds. This led, in 1971, to the banning of the tetracyclines and other 'therapeutic' antibiotics, in feed, as growth promoters. Bacitracin, virginiamycin, nitrovin and flavomycin, amongst others, were therefore promoted as safe antibiotics. However, their use may cause concern as they may affect the natural ecological balance in favour of pathogens such as *Salmonellae* (Smith, 1971). It has been proposed that the increase in the incidence of antibiotic resistant bacteria may be as a result of their over-use. In some countries, e.g. Sweden, the use of antibiotics as growth promoters is prohibited.

The mode of action of antibiotics

The theory of improved efficiency from the reduction in microbial loading by the use of anti-microbial agents centres on the fact that the gut is the most demanding of the body's organs both in terms of its energy and protein (Edmunds *et al.*, 1980; Murumatsu *et al.*, 1987). Any improvements in its nutritional efficiency will, therefore, have large effects on the animal's ability to improve its performance.

There has been much discussion about the mode of action of antibiotics which may have their effect through a variety of mechanisms, differing from one product to another. They have been shown to inhibit microbial growth, sugar metabolism, metabolite production, cell wall formation, and nucleic acid and protein synthesis.

Table 6.2 *Some modes of action of antibiotics in target bacteria.*

Mode of action (affecting)	*Antibiotics*
Bacterial cell wall	Penicillins, bacitracin
Bacterial cell membrane	Streptomycin group
Protein synthesis	Chloramphenicol, tetracyclines, lincomycin, neomycin, the streptomycin group
Nucleic acid metabolism	Griseofulvin
Intermediary metabolism	Sulphonamides

One possible mode of action is that antibiotics suppress some types of bacteria. This would give rise to the direct absorption of some materials which would be a more efficient use than their fermentation by microbes. The micro-flora normally metabolises some of the ingested food, with carbohydrates, for example, being broken down to lactic acid and volatile fatty acids under anaerobic conditions. These end products may be useful to the host after absorption, but this would depend on the site of fermentation. They are less useful and have less metabolisable energy than the original carbohydrates and, in the pig, for example, the major site of fermentation, the large intestine, is not regarded as an efficient site of absorption. Thus, antibiotics, if they act on the gut flora, may save energy by sparing carbohydrates from bacterial metabolism, reducing the production of volatile fatty acids, improving nitrogen availability by sparing essential amino-acids and reducing the levels of toxic amines (Henderick *et al.*, 1980). However, it is claimed that their growth promoting activity cannot be adequately explained solely by the suppression of some bacterial groups alone. Possible modes of action are summarised in Table 6.2 with a primary action possibly followed by a secondary anatomical or biochemical change (Feingold, 1963).

A classification of antibiotics may be based on whether their effect is bactericidal or bacteriostatic (Table 6.3) but such divisions are not always clear (Manten and Meyerman-Wisse, 1962).

Table 6.3 *Classification of antibiotics.*

Mode of action (affecting)	*Antibiotics*
Bacteriostatic	Tetracyclines, chloramphenicols, macrolides, sulphonamides
Bactericidal	Penicillins, novobiocin, bacitracin, polymyxins (including colistin), nitrofurans

Effects of antibiotics in the host animal

Numerous influences of antibiotics have been reported, many of which are interrelated.

First, harmful bacteria are inhibited. For example, *S. faecalis* has been shown to reduce growth rate in chickens (Eyssen and DeSomer, 1967), possibly by inducing malabsorption of fat with the influence of antibiotics improving weight gain and restoring fat absorption to normal. This may be due to nutrient sparing effects by inhibiting fermentation losses or by enhancing fat digestibility. (Fuller, 1984; Lev and Forbes, 1959; Smith, 1972; Stutz and Lawson, 1984).

A protein sparing effect is apparent through the reduction of faecal nitrogen loss which has resulted from antibiotic use (March *et al.*, 1978; Eyssen and Desomer, 1967; Visek, 1978). Furthermore, a sparing of amino-acids has been shown by Hedde (1984) who demonstrated that lysine was destroyed less in the gut of pigs when virginiamycin was included in the diet. This sparing effect has been observed as increased nitrogen retention in pigs (Lindsey *et al.*,1985). Such improvements are associated with a decreased passage rate of digesta and mucosal cell turnover. A slower rate of passage allows a greater time for digestion so improving nutrient availability. Consequently, there is improved nutrient absorption. This may also result from gut thinning (Gordon, 1952), which may partially account for improved absorption and increases in apparent digestibilities of protein and amino-acids (Ellis *et al.*, 1983). Feed antibiotics help to reverse the effects of the normal flora on the gut wall, which after treatment will resemble that of a germ free animal (Coates *et al.*, 1955).

Other likely effects of the use of anti-microbials include:

- Lower crypt cell proliferation leading to a reduction in protein turnover.
- Lower use of absorbed nutrients by microbial cells leaving increased quantities available to the animal's tissues.
- A reduction in epithelial turnover and mucus secretion conserving nitrogen.
- An increase in digestive and absorptive ability.

The use of antibiotics on the farm

The main production effects of using antibiotics in poultry and other animals are improved health performance, growth and feed utilisation. However, the Swann Report in 1969 first attempted to separate the use of anti-microbials into "in feed" and "therapeutic" types. The report also stated that the routine use of low levels of antibiotics "in feed" might endanger human health because of the development of bacterial resistance. Therapeutic antibiotics (which are used at much higher levels) for animal use are now only available on prescription in the United Kingdom. While up to 1989, in-feed antibiotics were sold without prescription, some are not now allowed for this purpose. Despite a bad press, linked with perceptions of them being 'unnatural', there are numerous reports to support the beneficial effects of antibiotics, for example in young calves (Brown *et al.*, 1960; Bush *et al.*, 1959; Lassiter, 1955; Lassiter *et al.*,

1958; Radisson *et al.*, 1956; Rusoff *et al.*, 1959; Swanson, 1963; Yates, 1962; Eyssen *et al.*, 1962).

The United Kingdom, like many other countries, has stringent regulations concerning materials that can be used in animal production. Materials that are added to the diet are divided into two categories. The Pharmaceutical Merchants List (PML) contains in Part B materials which have a product licence and may be used for incorporation into animal feeds. It contains such materials as coccidiostats, anthelminthics and some antibiotics. Prescription Only Medicines (POM) are those that may only be used when authorised by a veterinary surgeon's written direction. They include such materials as terramycin and oxytetracycline.

Furthermore, in animals destined for meat, periods before slaughter when antibiotics may not be used, are often prescribed. These withdrawal periods vary depending on the nature of the material and species of animal, etc. In the case of the various proprietary chlortetracyline products, a withdrawal period of up to 15 days would be common. On the other hand, there is no withdrawal period for zinc bacitracin.

These precautions are designed to ensure that the use of pharmacologically active substances does not result in residues in the livestock product. EEC legislation on residues is designed to give the consumer a measure of protection.

Effects from antibiotics on farm

A typical on farm antibiotic is avoparcin. It has been claimed to have activity in the gut but not to be absorbed into the body. Avoparcin is claimed to:

- Increase the villi surface area.
- Enhance nutrient absorption.
- Alter fermentation patterns in the rumen.
- Reduce Gram-positive organisms which damage villi and reduce nutrient supply.
- Maintain a thinner gut wall.
- Reduce epithelial cell turnover leading to increase in enzyme secretion (especially dipeptidase).
- Have a protein sparing effect.

Because of the better intestinal activity and nutrient absorption, it is claimed feed utilisation is improved in cattle, pigs and chickens, with milk yield improvement in dairy cattle.

Avoparcin is produced by a natural fermentation on a nutrient medium by a bacterium isolated from soil. It is included at low levels of 10 to 40 mg/kg.

In the ruminant animal improvements in performance, from in-feed antibiotics, eg. monensin sodium, are claimed to be due to alteration of rumen fermentation by inhibition of specific rumen micro-organisms. Fermentation in the rumen can therefore be restricted to more efficient and beneficial organisms (see Table 6.4).

Antibiotic resistance

One of the main concerns over the use of antibiotics for farm animals is the development of resistance, making them less effective. Most large bacterial populations contain

Table 6.4 *Sensitive and insensitive bacteria to monensin sodium in the rumen.*

Monensin sodium sensitive	*Products of fermentation*
Ruminococcus	Acetate
Methanobacterium	Acetate, Methane
Lactobacillus	Lactate
Butyrivibrio	Acetate, Butyrate
Streptococcus	Lactate
Methanosarcina	Methane
Selenomonas	Propionate
Bacteroides	Acetate, Propionate
Veillonella	Propionate

mutants which are naturally less susceptible to a given drug than the remainder of the population. Microbial resistance to anti-microbial agents may be due to physiological adaptation (phenotypic change) or to mutation and selection (genotypic changes). Once a drug-resistant mutant has emerged in a bacterial population, the resistance can be transferred to other cells by the mechanisms of transformation (incorporation of DNA from resistant organisms into sensitive organisms), transduction (transfer of drug resistance by bacteriophage), or conjugation (sexual reproduction and direct transfer).

Other suggestions for drug resistance include:

- Decreased permeability of the organism to the drug.
- Increased destruction of the drug.
- Conversion to an inactive form.
- Increased formation of the metabolites which compete with the drug.
- Production of inhibitory (extra cellular) enzymes.
- Development of alternate metabolic pathways.
- Altered enzymes which can function in the presence of the drug.
- Change in ribosomal protein structure.
- Duplication of target site.
- Reduction in physiological importance of site.

Transferable antibiotic resistance has been shown to be possible via plasmids called R-factors. The feeding of antibiotics is thought to favour the selection of R-plasmid bacteria in pigs and poultry, possibly leading to transfer of resistance to pathogenic bacteria such as *Salmonella*. The transferring of genes has been confirmed in Gram-positive bacteria, e.g. *Streptococcus*, *Bacillus*, *Staphylococcus* and *Clostridia*. R-plasmids are spread widely and found in the intestinal micro-flora of animals which have not undergone the pressure of antibiotic treatment. Cells can adapt, mutate or gain natural resistance to an antibiotic.

Contamination of animal products with antibiotics

Antibiotics have been claimed to leave residues in tissues (Wu, 1987), cause disturbance of other microbes, kill desirable bacteria, and cause pathogenic resistant strains (Wu, 1987). The level of antibiotic actually entering the animal's tissues will depend on the degree to which it is absorbed from the intestinal tract. Tetracyclines are generally well absorbed and so are more likely to enter the tissues than streptomycin and neomycin, which are less well absorbed.

In a United Kingdom survey of anti-microbial activity in meat and kidney samples of slaughtered pigs from 1980–83, by the MAFF Food Science Division (Ministry of Agriculture, Fisheries and Food, 1987), positive results for anti-microbial compounds were obtained in about 1% of samples, with residue levels of some antibiotics exceeding limits recommended by the Veterinary Products Committee.

The National Research Council of the USA (1980) also stated that if withdrawal periods (i.e. time from feeding to slaughter for metabolism of the antibiotic to reach a safe level) were adhered to, there would be insignificant levels remaining in the carcass. However, as regulations are not always observed, it has led to increased testing at abattoirs.

A proportion of the human population is sensitive to various antibiotics and it is generally thought that this is below 10% (Table 6.5). On the basis of long term microbiological studies, it has been suggested that the use of antibiotic growth promoters does not increase the prevalence of antibiotic resistance in humans (Lacey, 1988). Contrary to this belief other researchers have shown close relationships between drug resistance patterns in human and animal populations, with some reporters linking this to the utilisation of antibiotics as feed additives (Stabler *et al.*, 1982). Various suggestions to reduce the development of resistant strains have been made:

- The use of high strength antibiotics only until infection is overcome.
- The use of chemically unrelated drugs with different modes of action.
- Stronger control of anti-microbials in farming.
- The use of more 'natural' alternatives for therapeutic purposes and for animal growth and performance.

A full list of antibiotics permitted in livestock feed in Europe can be found in Table 6.6 (EEC Directive 70/524), detailing species, age and the minimum withdrawal period. Table 6.7 lists all the permitted additives under EEC Directive 70/524.

Table 6.5 *Sensitivity of the population to antibiotics (%).*

Antibiotic	*% of population*	*Source*
Penicillin	8	Katz (1980)
	7–10	Pace (1980)
Neomycin and streptomycin	5	Pace (1980)

Table 6.6 *Permitted Antibiotics in Europe (EEC Directive 70/524) as of 1st January 1994.*

EEC No.	Additive	Chemical formula description	Species or category of animal	Maximum age	Minimum Content (mg/kg of complete feedingstuffs)	Maximum Content (mg/kg of complete feedingstuffs)	Other provisions
	A. ANTIBIOTICS						
E 700	Bacitracin zinc	$C_{66}H_{103}O_{16}N_{17}S\ Zn$ (polypeptide containing 12 to 20% zinc)	Laying hens	–	15	100	–
			Turkeys	4 weeks	5	10	–
			Other poultry, excluding ducks, geese, pigeons	26 weeks	5	20	
						50	
				4 weeks	5		
				16 weeks	5	20	–
			Calves, lambs, kids	16 weeks	5	50	–
				6 months	5	20	–
				6 months	5	80	Milk replacers only
			Piglets	4 months	5	50	–
				3 months	5	80	Milk replacers only
			Pigs	6 months	5	20	–
			Animals bred for fur.	–	5	20	–
E710	Spiramycin	I $C_{43}H_{74}O_{14}N_2$ II $C_{45}H_{76}O_{15}N_2$ base III $C_{46}H_{78}O_{15}N_2$ (macrolide)	Turkeys	26 weeks	5	20	–

			Other poultry, excluding ducks, geese, laying hens, pigeons	16 weeks	5	20	–
			Calves, lambs, kids	16 weeks	5	50	–
				3 months	5	80	Milk replacers only
			Pigs	6 months	5	20	–
			Animals bred for fur.	–	5	20	–
E711	Virginiamycin	I $C_{28}H_{35}O7N_3$	Turkeys	26 weeks	5	20	–
		II $C_{43}H_{49}O_{10}N_7$	Other poultry, excluding ducks, geese, laying hens, pigeons.	16 weeks			
			Piglets	4 months	5	50	–
			Pigs	6 months	5	20	–
			Calves	16 weeks	5	50	–
				6 months	5	20	–
				6 months	5	80	Milk replacers only
			Laying hens	–	20	20	–
			Cattle for fattening	–	15	40	Indicate in the instruction for use: The quantity of virginiamycin in the daily ration must not exceed 140mg for 100kg of bodyweight and 6mg for each additional 10kg of bodyweight.

Table 6.6 *Contd.*

EEC No.	Additive	Chemical formula description	Species or category of animal	Maximum age	Minimum Content	Maximum Content	Other provisions
					mg/kg of complete feedingstuffs		
	A. ANTIBIOTICS						
E712	Flavophospholipol	$C_{70}H_{124}O_{40}N_6P$	Laying hens	–	2	5	–
			Turkeys	26 weeks	1	20	–
			Other poultry excluding ducks, geese, pigeons	16 weeks	1	20	–
			Piglets	3 months	10	25	Milk replacers only
			Pigs	6 months	1	20	–
			Animals bred for fur, excluding rabbits	–	2	4	–
			Calves	6 months	6	16	–
				6 months	8	16	Milk replacers only
			Cattle for fattening	–	2	10	Indicate in the instructions for use: The quantity of flavophospholipol in the daily ration must not exceed 40 mg for 100kg of bodyweight and 1.5mg for each additional 10kg bodyweight.
			Rabbits	–	2	4	–

E713	Tylosin phosphate	Macrolide, product of *Streptomyces fradiae* Composition of antibiotic	Piglets	4 months	10	40	–
		factors[1] (a) Tylosin $C_{46}H_{77}NO_{17}$: min 80% (b) Desmycosin $C_{39}H_{65}NO_{14}$ (c) Macrocin $C_{45}H_{75}NO_{17}$ (d) Relymycin $C_{46}H_{79}NO_{17}$ (a)+(b)+(c)+(d) : min 95%	Pigs	6 months	5	20	–
E714	Monensin sodium	$C_{36}H_{61}O_{11}Na$ (Sodium salt of a polyether monocarboxylic acid produced by *Streptomyces cinnamonensis*)	Cattle for fattening	–	10	40	Indicate in the instructions for use: The quantity of monensin sodium in the daily ration must not exceed 140mg for 100kg of bodyweight and 6mg for each additional 10kg of bodyweight. Dangerous for equines. This feeding stuff contains anionophore: simultaneous use with certain medicinal substances (eg tiamulin) can be contra-indicated.

[1]According to the method of analysis of the British Pharmacopoeia (Veterinary) 1985.

Table 6.6 *Contd.*

EEC No.	*Additive*	*Chemical formula description*	*Species or category of animal*	*Maximum age*	*Minimum Content* mg/kg of complete feedingstuffs	*Maximum Content* mg/kg of complete feedingstuffs	*Other provisions*
	A. ANTIBIOTICS						
E716	Salinomycin sodium	$C_{42}H_{69}O_{11}Na$ (sodium salt of a polyether monocarboxylic acid produced by *Streptomyces albus*)	Piglets	4 months	30	60	Indicate in instructions for use: Dangerous for equines. This feedingstuff contains an ionophore simultaneous use with certain medicinal substances (eg tiamulin) can be contra-indicated.
			Pigs	6 months	15	30	
E715	Avoparcin (glycopeptide)	$C_{53}H_6O_{30}N_6Cl_3$	Chickens for fattening				–
			Turkeys for fattening	16 weeks	10	20	–
			Piglets	4 months	10	40	–
			Pigs	6 months	5	20	–
			Calves	6 months	15	40	–
			Cattle for fattening	–	15	30	Indicate in the instructions for use: The quantity of avoparcin in the daily

							ration must not exceed 100mg for 100kg of bodyweight and 4.3mg for each additional 10kg of bodyweight
			Lambs from the beginning of rumination, with the exception of pasture-grazed lambs	16 weeks	10	20	
E717	Avilamycin	$C_{57-62}H_{82-90}$ $Cl-_{1-2}O_{31-32}$ (mixtures of oligosaccharides of the orthosomycin group produced by *Streptomyces viridochromogenes*)	Piglets	4 months	20	40	–
			Pigs	6 months	10	20	–
E22	Avorpacin	$C_{53}H_6O_{30}N_6Cl$ (glycopeptide)	Lambs from the beginning of rumination, with the exception of pasture-grazed lambs	16 weeks	10	20	
			Dairy cattle	–	4	10	Indicate in the instructions for use: The quantity of avoparcin in the daily ration must not exceed 100mg and for reasons of efficacy, must not be less than 50mg

Table 6.6 *Contd.*

EEC No.	Additive	Chemical formula description	Species or category of animal	Maximum age	Minimum Content mg/kg of complete feedingstuffs	Maximum Content mg/kg of complete feedingstuffs	Other provisions
	A. ANTIBIOTICS						
E28	Avilamycin	$C_{57-62}H_{82-90}Cl_{1-2}O_{31-32}$ (Mixtures of oligosaccharides of the orthosomycin group produced by *Streptomyces viridochromogenes*)	Chickens for fattening		2.5	10	–
E29	Efrotomycin	$C_{59}H_{88}N_2O_{20}$	Piglets	4 months	4	8	–
			Pigs	6 months	4	6	–
E30	Virginiamycin	$1C_{28}H_{35}O_7N_3$ $11C_{43}H_{49}O_{10}N_7$	Sows	–	20	40	–

Table 6.7 *Permitted feed additives under EEC Directive 70/524 as of 1st January 1994.*

Antibiotics	*Coccidiostats and other medicinal substances*
AvilamycinAmprolium	Amprolium/ethopabate
Avoparcin	Decoquinate
Bacitractin zinc	Diclazuril
Efrotomycin	Dimetridazole
Flabophospholipol	Dinitolmide
Monensin sodium	Halofuginone
Salinomycin sodium	Ipronidazole
Spiramycin	Lasalocid sodium
Tylosin phosphate	Meticlorpindol
Virginiamycin	Meticlorpindol/methylbenzoquate
	Monensin sodium
	Narasin
	Nicarbazin
	Nifursol
	Robenidine
	Ronidazole
	Salinomycin sodium
Trace Elements	
Copper	
Zinc	
Growth Promoters	
Carbadox	
Olaquindox	

New developments are resulting in antibiotics which are more selective in their action and, in some cases, the new antibiotics have little effect on the populations of beneficial, non-pathogenic micro-organisms within the gut environment. It has been suggested that it is desirable to find more natural methods of regulating the number of pathogenic bacteria, other than by the routine use of antibiotics. Antibiotics are, however, undoubtedly essential in modern agricultural systems and their importance should not be underestimated.

CHAPTER 7

Lactic acid bacteria

Introduction

Special attention is given to the lactic acid bacteria because of their importance in the nutrition of animals and man. In nature, their occurrence is related to their high demands for nutrients and to the fact that their energy generation is purely by fermentation. They are rarely found in soil or water. Examples of their occurrence are:

- In milk, its products (e.g. yoghurt) and the places where it is processed with the presence of, e.g. *Lactobacillus lactis, L. bulgaricus, L. helveticus, L. casei, L. fermentum, L. brevis, Streptotoccus lactis.*
- In the mucous membranes and the gastro-intestinal tract of animals, there are, e.g. *Lactobacillus acidophilus, Bifidobacterium, Streptotoccus faecalis, S. salivarius, S. bovis, S. pyogenes, S. pneumoniae.*
- In both rotting (e.g. *Lactobacillus plantarum, L. delbruckii, L. fermentum, L. brevis, Streptococcus lactis, Leuconostoc mesenteroides*) and intact plants, e.g. in silage preservation, where they are present naturally and are also introduced to suppress the growth of harmful bacteria.

Streptococcus (also known as *Enterococcus*) *faecalis* is a normal inhabitant of human intestines with many *Streptococci* found on the mucous membranes of the oral cavity, and the respiratory, urinary and genital organs, while some are blood parasites and very virulent pathogens.

Lactic acid is the main acid in sour milk. It was first identified as a fermentation product in 1847 by Blondeau, and in 1877, a pure culture of lactic acid producing bacteria, *Streptococcus lactis*, was isolated. Feeding lactic acid was shown to lead to improved growth in weaned pigs and a shift in the bacterial population against *E. coli* (Cole *et al.*, 1968), and other work supports this phenomenon (Herrick, 1972; Pollman *et al.*, 1980; Wu, 1987). Lactic acid has also been found to be inhibitory towards *S. typhimurium, in vitro* (Rubin and Vaughan, 1979), and this is of obvious importance in the control of *Salmonella* on a routine basis.

Lactic acid bacteria

Man has known of the ability of lactic acid bacteria to ferment substrates for over one hundred years (i.e. they were known to cause fermentation and coagulation of milk). Weigmann (1899a) produced the first definition of the lactic acid bacteria as those which produce milk acid (lactic acid) from milk sugar (lactose).

The original subdivision of the genuine lactic acid bacteria was into six genera; *Betacoccus, Streptococcus, Tetracoccus, Betabacterium, Streptobacterium* and *Thermobacterium* (Davis, 1960). The genus names *Betacoccus and Tetracoccus* have been replaced by *Leuconostoc* and *Pediococcus*. These same classifications are recognised today (Bergey, 1986). A further classification has resulted in two subdivisions depending on whether they are able to ferment glucose solely to lactate or to other products as well, i.e. homo-fermentative or hetero-fermentative (Jensen, 1943). (See Table 7.1).

Homo-fermentative

Homo-fermentative lactic acid bacteria metabolise glucose via the fructose-bisphosphate pathway, (see *Figure 9.12*) having all the necessary enzymes, including aldolase, and being able to use the hydrogen obtained from the dehydrogenation of glyceraldehyde-3–phosphate (to 1,3–bisphosphoglycerate) to reduce pyruvate to lactate.

In this case, lactate, which is the product of glucose fermentation, can account for up to 90% of the end products. The extent to which other products occur, depends on oxygen supply.

Hetero-fermentative

In the case of these bacteria, fifty per cent of the end products of glucose metabolism is lactic acid. In addition, large amounts of CO_2 (20–25%), acetic acid and ethanol are also produced. Mannitol is obtained from fructose in this sub-group and other determinative points include:

- Gas produced during fermentation of glucose and gluconate.
- Fermentation of ribose to lactic acid without gas production.
- Thiamine is required for growth.
- Glucose-6–phosphate dehydrogenase activity is shown.

Examples of lactic acid bacteria in this group are:

(i) *L. fermentum*. Growth at 45°C but none at 15°C.
(ii) *L. cellobiosus*. Variable growth at 45°C and at 15°C, but no growth at 48°C.
(iii) *L. viridescens*. Growth at 15°C, but none at 45°C.

Table 7.1 *Lactic acid bacteria classification depending on shape (rods or cocci) and type of fermentation.*

Cocci	*Rods*
Fermentation pathway	
HOMO-FERMENTATIVE $C_6H_{12}O_6 \longrightarrow 2CH_3 - CHOH\text{-}COOH$	
Streptococcus cremoris	Thermobacteria
Streptococcus diacetilactis	*Lactabacillus lactis*
Streptococcus faecalis	*Lactabacillus helveticus*
Streptococcus lactis	*Lactabacillus acidophilus*
Streptococcus pyogenes	*Lactabacillus bulgaricus*
Streptococcus salivarius	*Lactabacillus delbruckii*
Streptococcus thermophilus	Streptobacteria
Pediococcus cerevisiae	*Lactobacillus casei*
Lactobacillus plantarum	
Sporalactobacillus inulinus	
Fermentation pathway	
HETERO-FERMENTATIVE $C_6H_{12}O_6 \longrightarrow CH_3 - CHOH\text{-}COOH + CH_3 - CH2OH + CO_2$ (or $CH_3\text{–}COOH$)	
Leuconostoc mesenteroides	Betabacteria
Leuconostoc cremoris	*Lactobacillus brevis*
	Lactobacillus fermentum
	Lactobacillus viridescens
	Bifidobacterium bifidum

(*Based on Schegel, 1986*)

The species classification can be confirmed by other complex methods, apart from temperature of growth, which include the chemical nature of the cell wall, cell wall membrane, antigenic determination, serological grouping (Sharpe, 1970; Knox and Wickes, 1973), vitamin requirement (Rogosa *et al.*, 1961), amino acid sequences of the peptidoglycan and DNA composition (Gasser and Mandel, 1968) and DNA homology (Simmonds *et al.*, 1971; Dellaglio *et al.*, 1973; 1975).

The hetero-fermentative lactic acid bacteria lack the important enzymes of the fructose-bisphosphate pathway, aldolase and triose-phosphate isomerase and details of the pathway are given in *Figure* 9.13.

Hetero-fermentative bacteria can convert the acetylphosphate, either partially or completely, to acetate gaining utilisable energy as ATP. They ferment fructose with the formation of lactate, acetate, carbon dioxide and mannitol:

$$3 \text{ fructose} \longrightarrow \text{lactate} + \text{acetate} + CO_2 + 2 \text{ mannitol.}$$

The fructose can also accept excess reducing equivalents e.g.

$$\text{fructose} + \text{NADH}_2 \longrightarrow \text{mannitol} + \text{NAD}$$

Lactobacilli

Lactobacilli are one of the most dominant groups of bacteria found in parts of the gastro-intestinal tract. Classically, the cells are large, and appear as pairs or short chains. They are non-motile, non-sporing and non-capsulate, grow over a wide temperature range (15–45°C), but have an optimum of 37°C. They prefer acidic conditions (pH 5.8). They have been shown to increase in number from birth, whereas other bacteria decline (Fuller, 1989). Common *Lactobacilli* in the intestinal flora include *L. acidophilus*, *L. bifidus*, *L. leichmannii*, *L. plantarum*, *L. casei* and *L. fermentum*.

Lactobacilli are known to produce bacteriocins (antibiotic-like compounds) that can attack other bacteria. Shahani and Ayebo (1980), isolated two of these: acidophilin from *Lactobacillus acidophilus* and bulgarican from *Lactobacillus bulgaricus*.

Lactobacilli have an ability that most micro-organisms lack, i.e. to utilise lactose. This ability is shared with a number of intestinal bacteria (e.g. *E. coli*). Lactose is produced only by the mammary gland of mammals and is secreted in milk which is ingested with the utilisation of lactose by micro-organisms. Lactose is a disaccharide that must be hydrolysed before it can enter the catabolic pathway for hexoses.

$$\text{lactose} + H_2O \xrightarrow{\beta\text{–galactosidase}} \text{D-glucose} + \text{D-galactose}$$

The galactose is then phosphorylated and converted to glucose phosphate.

Bifidobacteria

The hetero-fermentative lactic acid bacterium called *Bifidobacterium bifidum* takes its name from the Y-shape of its cells (Latin *bifidus*, divided into two). It is dominant in the intestinal tract of babies, especially those that are breast fed (Hall *et al.*, 1990; Benno and Mitsuoka, 1986). This specificity of their distribution to breastfed babies can be traced to the high requirement of this bacterium for sugars that contain N-acetylglucosamine, which is found only in human milk and not, for example, in cow's milk. The members of the genus *Bifidobacterium* are strict anaerobes.

Lactic acid bacteria in agricultural and food production

If non-sterile solutions containing sugars, complex nitrogen sources, and accessory factors are left under anaerobic conditions they will soon become overgrown with lactic acid bacteria. These lower the pH to below 5 and suppress the growth of other anaerobic bacteria, which are less acid tolerant. The actual type(s) of lactic acid bacteria that become dominant in these cultures depends on specific conditions. This sterilising and preserving effect of the lactic acid bacteria, due to their acid production, has led to their use in agriculture (e.g. silage making) and food processing (e.g. yoghurt and other milk-utilising products).

Silage production

Silage is an important cattle feed, generally made from grass or other forage crops which are high in moisture. However, other materials, such as fish, can be preserved in this way. The objective in all silage manufacture is to produce sufficient acid to prevent the activities of spoilage organisms such as *Clostridia*. If, however, the activity of *Clostridia* is not prevented then butyric acid is produced, protein is broken down and feeding value is reduced. The pH required will depend on the conditions, but normally pH4 is satisfactory. Consequently efforts are made to aid the processes of silage making. Various aids are used, e.g. molasses, enzymes, acids and bacterial inoculants (see Table 7.2).

The high moisture and physical compaction of forages exclude air and prevent spoilage by microbes which thrive in air. Acid fermentation occurs, in which the lactic acid bacteria present in the ensiled material ferment the sugars to lactic and other acids.

Table 7.2 *Current methods used to manipulate silage fermentation and preservation.*

Application	*Objective*	*Problems/concerns*	*Potential benefits*
Molasses	Supply simple sugars to feed lactic acid bacteria and thereby reduce pH due to lactic acid produced	Difficulty of application. Can stimulate less efficient heterofermenters	Cost effective. Provides residual energy and protein as well as improved fermentation
Acids	Lower pH of ensiled material to stabilise/ preserve silage	Can be corrosive and lock up natural minerals	No use of crop nutrients in unneccessary fermentation
Enzymes	Convert fibrous carbohydrates to sugars and allow naturally present lactic acid bacteria to lower pH	Activity needs careful controlling. Can be relatively expensive compared with other additives	Can work on higher dry matter and more fibrous crops. Natural crop sugars are not essential to feed lactic acid bacteria
Bacterial Innoculants	Produce maximum amount of lactic acid from available soluble carbohydrates (sugars). Hence pH is reduced and conditions produce a stable silage	Viability and strain benefits need controlling. Adequate numbers must be supplied	Homolactic fermentation improves efficiency of lactic acid production and fermentation

It has long been recognised that bacteria are involved in silage making, although it was originally thought that yeasts were of major significance. It is now known that yeasts play only a minor part in fermentation but are one of the main causes of loss during aerobic deterioration.

The main organisms playing a significant role in the ensilage process have been described (Woolford, 1984) as:

- The lactic acid-producing bacteria (*Lactobacilli*, *Streptococci*, *Leuconostoc* and *Pediococcus*).
- The endospore forming bacteria (*Clostridia* generally associated with spoilage) and bacilli.
- Fungi (yeasts and filamentous fungi) which are associated with spoilage.
- Other groups such as propionic acid bacteria (which may increase with the greater production of lactic acid) and *Listeriae* are occasionally found.

Microbial change during ensilage

In normal silage making there are two distinct stages of microbial development.

- Lactic acid bacterial development (always takes place).
- *Clostridial* development (may or may not take place).

The later may or may not occur, depending on the conditions prevailing, e.g. dry matter and sugar content.

Providing anaerobic conditions have been established after harvesting, there will be a shift in the microbial population from the aerobes of the fresh crop to facultative and obligate anaerobes of the silage. This is usually associated with a change from Gram-negative to Gram-positive organisms (Woolford, 1984).

Coliform bacteria multiply until about the seventh day after harvesting (Beck, 1972) and then decrease. At the same time they are replaced by lactic acid cocci including *Streptococci*, *Leuconostoc* and *Pediococci*, and finally by the higher yielding acid producing *Lactobacilli*. It is essential to control coliforms in the first seven days if quality is to be maintained.

Of the early appearing lactic acid bacteria, *Pediococci* dominate with *Lactobacilli* in the fermented silage (*Figure 7.1*). Total numbers of viable organisms following ensilage are usually of the order of 10^9 to 10^{10} per gram fresh weight, with the majority being lactic acid bacteria. This dominance is due to their powers of survival (tolerance to acid) rather than their ability to grow more rapidly than another species (Gibson and Stirling, 1959). *Pediococci*, for example, tend to be more acid tolerant and consequently maintain higher numbers than some other strains which are acid limited, e.g. *Streptococcus lactis*. Also, the lactic acid bacteria may produce other substances which are inhibitory to other native grass organisms. As the type of fermentation exhibited by the lactic acid bacteria changes, (within four days of ensilage) there is a move from homo-fermentative to hetero-fermentative fermentation again possibly due to the change in pH and increase in acetic acid.

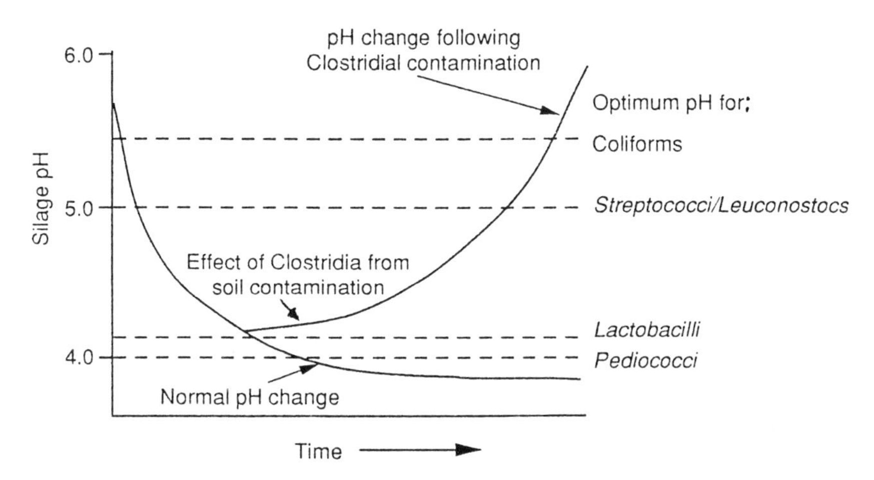

Figure 7.1 Qualitative changes which take place in the silage micro-flora during fermentation ---- = preferred pH of silage organism (after Woolford, 1984).

Figure 7.1 shows how the pH of the silage drops progessively with time until a stable crop of pH4 is produced (about day 10). The figure also indicates the pH optima of different organisms showing which would dominate at a given pH. It also shows how, when *Clostridia* are present (from soil contamination), the pH can rise, making the silage more suitable for coliforms. This rise in pH can be associated with an increase in proteolytic enzymes, which breakdown protein to ammonia and free amides.

Figure 7.2 illustrates the overall changes in nutrients from ensiling a fresh crop to stabilisation of a preserved crop. There are many different fermentation pathways involved within this general summary. For example, the fermentation of glucose is affected by whether the organisms are homolactic or heterolactic fermenters. Homolactic fermenters produce twice as much lactate and are therefore more efficient in producing conditions to preserve the crop (*Figure 7.3*).

Clostridia although often associated with the later stages of ensilage may also multiply in the first few days of ensilage. After this, their viable count decreases (Gibson *et al.*, 1958). If, howevei lactic acid bacteria ferment silage carbohydrates to butyric acid (causing spoilage), these conditions are conducive to the multiplication of proteolytic *Clostridia*.

The process can, therefore, be effective in the preservation of the ensiled crop, or, under adverse conditions, produce poorly stabilised silage of low feed value, from which valuable nutrients have been lost due to poor fermentation. Today, lactic acid bacteria play a critical role in this process (especially homolactic fermenters) and are consequently the content of many commercial silage additive products.

Silage inoculants

The theory of commercial inoculants centres around grass being a relatively poor source of lactic acid bacteria, especially those which are efficient convertors of silage dry matter to fermentation acids (homo-fermentative). After ensilage, their numbers can be low

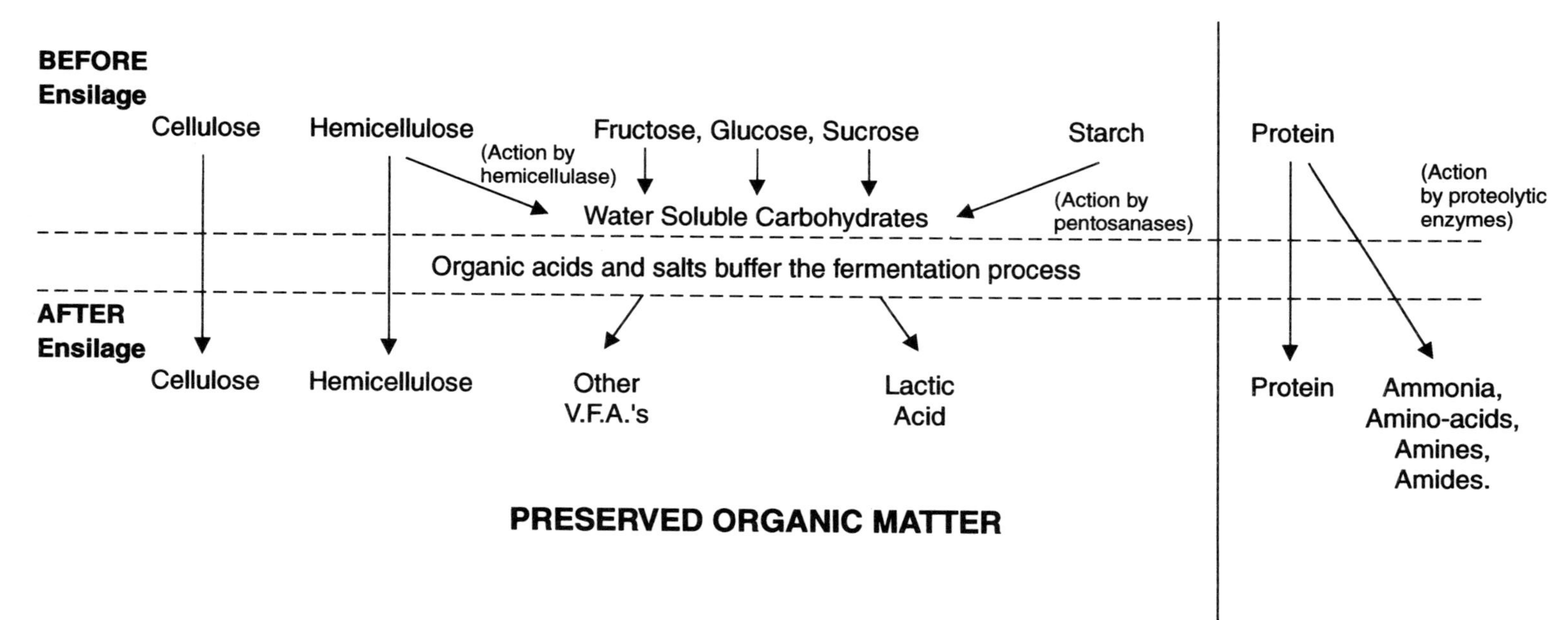

Figure 7.2 The effect on crop nutrients of ensiling of organic matter.

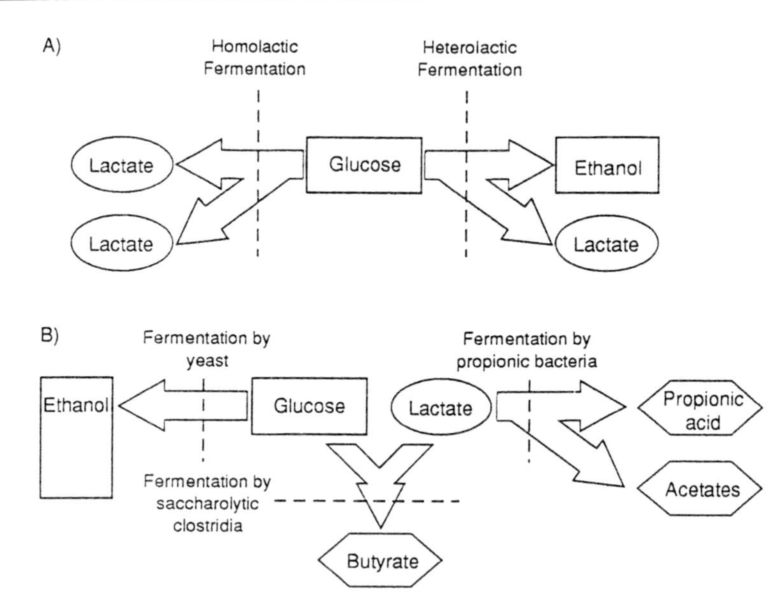

Figure 7.3 Fermentation products of specific silage organisms.
A) The basic fermentation products of carbohydrate (glucose) by lactic acid bacteria.
B) Potential end products of carbohydrate fermentation by silage spoilage organisms.

allowing the coliforms or enterobacteriaceae to dominate, producing acetic acid and ammonia. In order to overcome this it is claimed that the correct bacteria should be present, i.e. homo-fermentative organisms which convert natural sugars to lactic acid. The beneficial organisms used include *Streptococci*, *Pediococci*, *Lactobacilli*. A range of organisms is often used to provide organisms with different pH tolerances (see *Figure* 7.1). The stabilisation sequence would start with organisms such as *Streptococcus* being dominant. Finally *Lactobacilli*, which have a lower pH tolerance, take over. The addition of organisms leads to an increase in the total numbers of lactic acid bacteria in the crop (*Figure* 7.4) and allows a faster rate of lactic acid production (*Figure* 7.5). This enables the silage to reach peak lactic acid levels in fewer days, ensuring less nutrients are lost, and limiting the chances of challenge by spoilage organisms.

Fermented milk products

Man has theorised for years about the many beneficial effects of yoghurt in the human diet. It has been claimed to increase life-span, and on a day-to-day basis maintain the balance of bacteria and "settle" the environment within the digestive tract. Yoghurt has it origins stretching back to before Egyptian times when it was known as 'Benraib'. It was made by allowing the sugars in milk to ferment, using its natural bacteria. As time now shows, these bacteria were mainly lactic acid bacteria and they converted the lactose of milk to lactic acid. Late in the sixteenth century, Francis I of France was reported to have enjoyed the benefits of yoghurt. A persistent intestinal disorder could not be cured, but under advice from the Sultan of the Ottoman Empire, yoghurt was prescribed and the

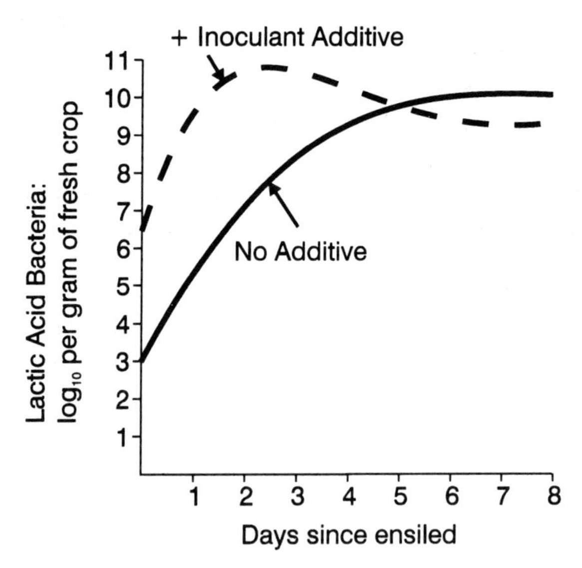

Figure 7.4 Lactic acid bacteria levels in control and inoculant treated silage in the first seven days of ensilage. Thomas and Slater (1986).

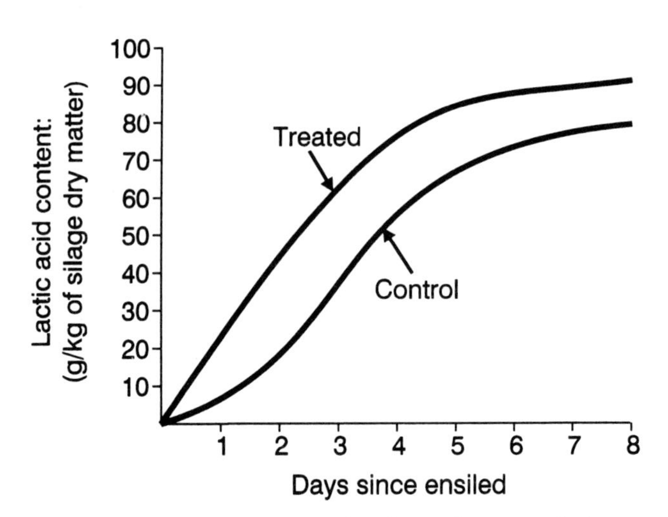

Figure 7.5 Lactic acid bacteria levels in control and inoculant treated silage in the first seven days of ensilage. (Thomas and Slater, 1986).

problem disappeared. Other fermentations of *Lactobacilli* have been equally important in the production of bread and beer.

In 1908, Metchnikoff won the Nobel prize for his work on yoghurt at the Pasteur Institute in Paris. He concluded that longevity was a result of maintaining a healthy intestine. This was accomplished by eating the correct diet. His research subjects were the Balkan mountain people and people of middle eastern countries who relied on yoghurt containing live bacteria for a large part of their diet.

Lactose is a simple carbohydrate present in milk and provides the energy necessary for the fermentation of milk. Lactic acid is a product of its fermentation and may be responsible for some of the beneficial effects of yoghurt. Other beneficial effects of yoghurt appear to be related to the functions of the live bacteria in the digestive tract. It is claimed, unlike most other ingested bacteria, that those in yoghurt can survive digestion in the stomach and small intestine, where they produce their effects on the native intestinal micro-flora.

The bacteria used in yoghurt are not major inhabitants of the intestinal flora of humans, but help to maintain the balance of this endogenous flora. Work on babies has shown that those fed live yoghurt had a significantly lower susceptibility to intestinal infections than babies fed on cow's milk (Larue, 1960; Mayer, 1962). Later work by Niv (1963) even found yoghurt to reduce diarrhoea in children more quickly than antibiotics, but antibiotics were still more effective at removing persistent long term problems. *Bifidobacteria* have been implicated as being important in the digestive process by maintaining a healthy balance of both good and bad bacteria (Lubis, 1983).

The first industrial yoghurt manufacturing process was developed in 1919 by Dr Carasso in Barcelona, using bacterial cultures derived from those of the Pasteur Institute.

Manufacture of yoghurt

There are six main factors necessary for the correct fermentation of milk by bacteria (see also *Figure* 7.6).

- Specific and selected live bacteria must be added.
- The temperature of the milk must be correct for fermentation (45°C).
- The bacteria must replicate by growing on the lactose present in the milk for 2–3 hours.
- No foreign organisms must be present. The milk is therefore pasteurised to destroy any harmful pathogens.
- As in other commercial fermentations, light agitation must be applied frequently.
- When fermented the medium must be chilled to prevent further secondary fermentation (4°C).

There are normally two types of bacteria used in the manufacture of yoghurt.

- *Streptococcus thermophilus.*
- *Lactobacillus bulgaricus.*

Commercially, various forms of milk and cream are inoculated with cultures of lactic acid producing bacteria, which act as 'starters' for fermentation. Examples of such products are given in Table 7.3. Again these products are a way of storing materials through the production of lactic acid, which lowers pH and suppresses other anaerobic, but harmful bacteria. Sour milk products are also used in cases of milk allergies and lactose intolerance.

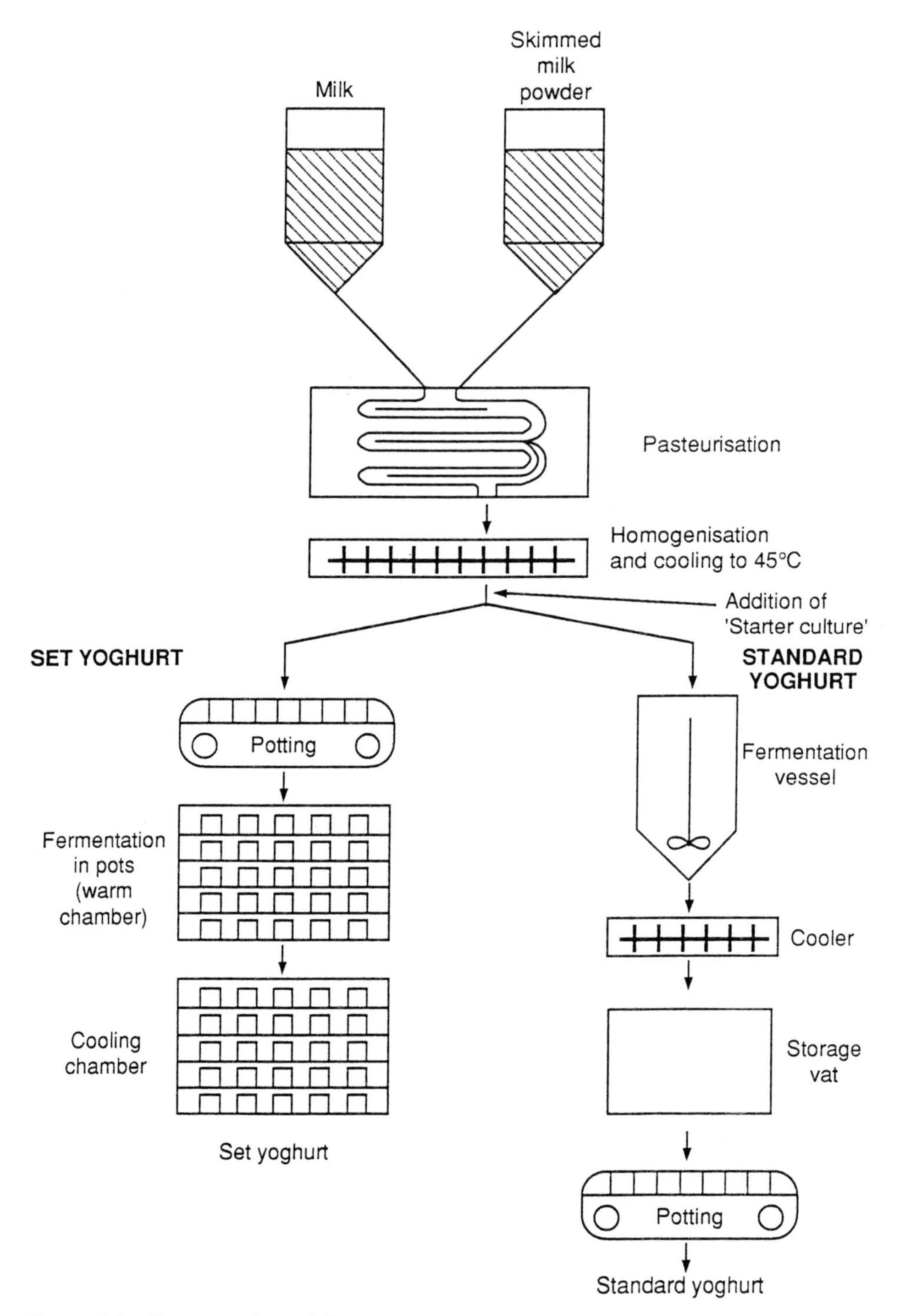

Figure 7.6 Illustration shows difference in process between standard yoghurt (pourable) and set yoghurt (non-pourable).

Table 7.3 *Examples of sour milk products.*

Product	*Culture*	*Temperature and time of incubation*
Sour cream and buttermilk	*Streptococcus lactis, S.cremoris, Leuconostoc cremoris or S. diacetilactis*	22°C, 18h
Yoghurt	*Streptococcus thermophilus, Lactobacillus bulgaricus*	43–45°C, 2.5–3h
Cottage cheese curd	*Streptococcus lactis, S.cremoris, Leuconostoc cremoris*	22°C, 18h or 35°C, 5h
Kefir	*Streptococcus, Lactobacillus* and yeast	15–22°C, 24–36h
Kumiss (made from donkey milk)	*Lactobacillus bulgaricus* and Torula yeast	15–22°C, 24–36h

(Partly based on Schlegel H.G., 1986)

Yoghurt has become particularly important commercially with the development, not only of many flavours but also of many variations, e.g. 'low fat', 'creamy', and 'set'. The pasteurised, homogenised whole milk is inoculated with *Streptotoccus thermophilus* and *Lactobacillus bulgaricus* to make yoghurt. An interesting feature is that there is a synergistic effect between these two organisms (*Figure* 7.7). On the one hand *L. bulgaricus* produces free amino acids, particularly histidine, which encourages the growth of *S. thermophilus* and this, in turn, produces formic acid-like compounds and CO_2 which stimulate the growth of *L. bulgaricus* (Tamine and Deeth, 1980).

The use of fermented milk products in the human diet to treat gastro-intestinal disorders is by modifying the gut flora to the detriment of the harmful bacteria. It is

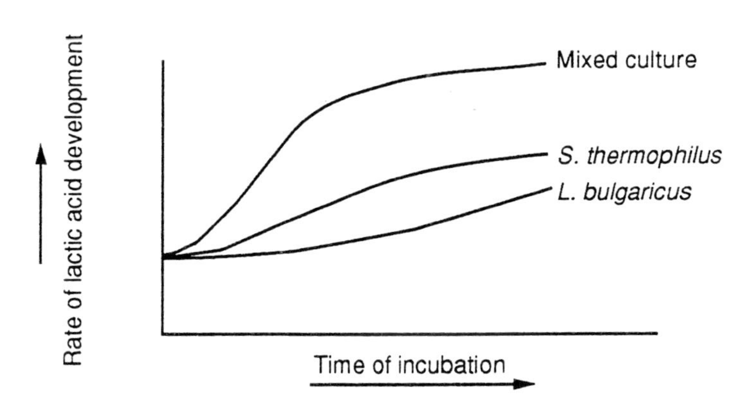

Figure 7.7 The lactic acid producing effects of *S. thermophilus* and *L. bulgaricus* when fed singly or in combination (based on Tamine and Robinson, 1988).

known that *L. bulgaricus* is able to survive through the gastro-intestinal tract of infants and also to inhibit some Enterobacteriaceae. When added to human milk in the diet of infants aged between one and four months, it has been claimed to improve weight gains. In the adult, a large increase in the numbers of *Lactobacilli* recovered in the faeces after yoghurt had been included in the diet for one or two weeks has been reported. Three weeks after yoghurt was removed from the diet the faecal flora had returned to its pre-treatment state. Initially there was a ten-fold reduction in coliform count, but after a further week, it returned to the pre-treatment level. An interesting feature of this work was that although there was a large increase in *Lactobacillus* count, they were not the same strains as contained in the yoghurt (W.M. Waites and E. Tibble, personal communication). It was suggested that the reasons for such a phenomenon may include a failure of *L. bulgaricus* to colonise and survive the large intestine, and also that it was a poor competitor with the indigenous species. Even the feeding of unfermented milk containing *L. acidophilus* to humans has resulted in a significant increase in *Lactobacilli* in the faeces.

Beneficial effects and possible modes of action of desirable gut micro-organisms in the gastro-intestinal tract

A number of suggestions have been put forward to explain how the beneficial micro-organisms might help promote health and combat the proliferation of pathogenic species of bacteria, and at present it is unclear which of these might be the most important. It is possible that several factors together might be involved.

Competitive exclusion

This theory has received considerable attention and proposes the prevention of colonisation of some micro-organisms (including pathogens) by others. In its simplest form it can be regarded as the beneficial bacteria occupying adhesion sites in the gut that would otherwise be populated by harmful bacteria. The complex interaction of good and bad micro-organisms protecting the animal has been termed 'bacterial antagonism' (Freter, 1956), 'bacterial interference' (Dubos *et al.*, 1963), 'barrier effect' (Ducluzeau, *et al.*, 1970), as well as 'competitive exclusion' (Lloyd *et al.*, 1977). The precise mechanism controlling the principle of gut colonisation is not well known but involves specific recognition of receptor sites (oligosaccharides) by bacterial fimbrae (lectins). *Lactobacilli* that colonise the stomach and associate with the non-secreting epithelium have been considered to be important in preventing populations of coliform bacteria (e.g. *E. coli* and its pathogenic and non-pathogenic relatives (Adler and Da Massa, 1980) from rising to high numbers. They are normal inhabitants of the large intestine and it is when they become prolific in the small intestine that problems occur for example, in pigs (Barrow *et al.*, 1977).

In poultry, significant protection against *Salmonella* has also been found by administration of intestinal micro-flora from selected donor birds (Snoeyenbos *et al.*, 1978; Vincent *et al.*, 1955; Luckey 1963; Weinack *et al.*, 1981; Muralidhara *et al.*, 1977;

Savage 1969; Shahani and Ayego, 1980). In competitive exclusion, *Lactobacilli* attach to the wall of the crop and compete with *E. coli* (Fuller and Brooker, 1974), *Salmonella* and other pathogens. These observations add support to the hypothesis suggested earlier by several investigators that 'competitive exclusion' helps prevent attachment of the pathogens to the intestinal mucosa. It has also been reported that newly hatched birds treated with the protective micro-flora, had partial protection within a few hours, and reached full potential in about 32 hours (Soerjadi *et al.*, 1981).

Competitive exclusion (*Figure* 7.8) has been reported to stop gut colonisation by *Salmonella* when the contamination level is very high. *Salmonella* excretion levels of chicks treated with probiotics, have been shown to be at least two logs lower than untreated chicks (Weinack *et al.*,1979). Furthermore, chicks which came into contact with *Salmonella* following treatment with normal gut micro-flora had reduced colonization by pathogens and toxin production was not seen (Snoeyenbos *et al.*, 1984). Meynell (1963) and Bonhoff *et al.* (1964) reported that volatile fatty acids (VFA) were produced in the caeca of mice by bacteria and linked the VFA produced with the protection against *Salmonella*. A molecular mechanism for attachment and invasion of epithelial cells by *Salmonella* has recently been reported by Finley *et al.*, (1989). Other mechanisms include competition for nutrients, reducing the redox potential, and non-specific activation of the immune system (Hill *et al.*, 1986).

Competitive exclusion can take place throughout the digestive tract as non-pathogenic organisms such as lactic acid-producing bacteria are present in the food. The lactic

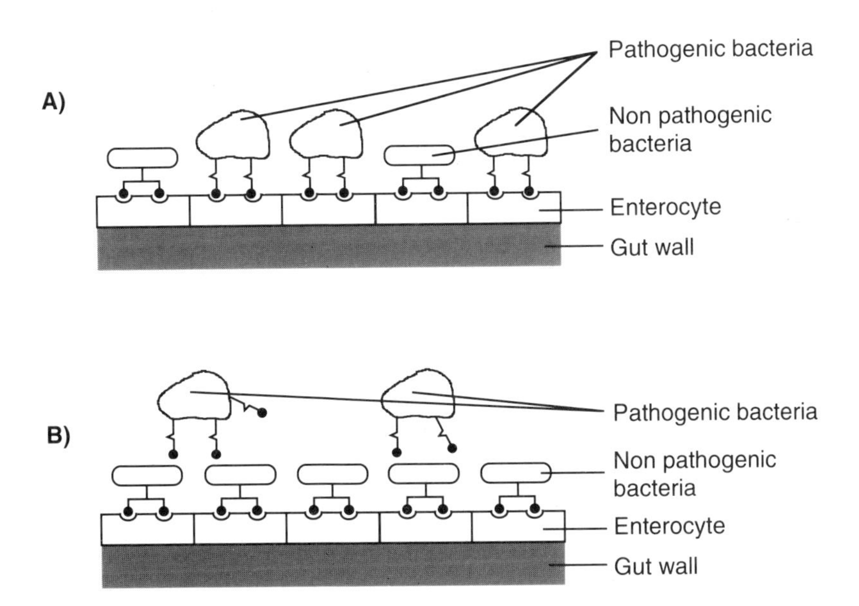

Figure 7.8 A) represents a mixed population of bacteria with a substantial attachment of pathogenic bacteria. B) shows competitive exclusion of pathogens due to preferential attachment of non-pathogens. It should be noted that the recognition of receptor sites (oligosaccharides) by the bacterial fimbrae (lectins) is very specific to different types of organisms (see Chapter 4).

acid-producing bacteria flow from the stomach into the small intestine where their lactic acid production may mediate, at least in part, the inhibition of the development of pathogenic Gram-negative bacteria. *L. acidophilus* has been shown to be effective in adhering to the intestinal epithelium of the pig (Conway *et al.*, 1987), and it is agreed (Jones and Rutter 1972; Linton and Hinton, 1988) that *E. coli* colonisation is necessary for its pathogenic activity. This is further supported by observations of reduced numbers of *Lactobacilli* and an increase in *E. coli* colonising the intestine walls in scouring pigs (i.e. suffering from diarrhoea) compared with their healthy counterparts (Muralidhara *et al.*, 1973). Intestinal material has been used to provide bacteria to block *Salmonella* from attachment sites (Pivnick and Nurmi, 1982; Mead and Impey, 1987). Large scale field trials looking at competitive exclusion have also reportedly produced good results, (Mead and Impey, 1989).

It has also been reported that *Lactobacilli* can successfully dominate other bacteria in the competition for nutrients in the gut and therefore survive to colonise the intestine (Muralidhara *et al.*, 1977; Roach *et al.*, 1977) (*Figure 7.9*).

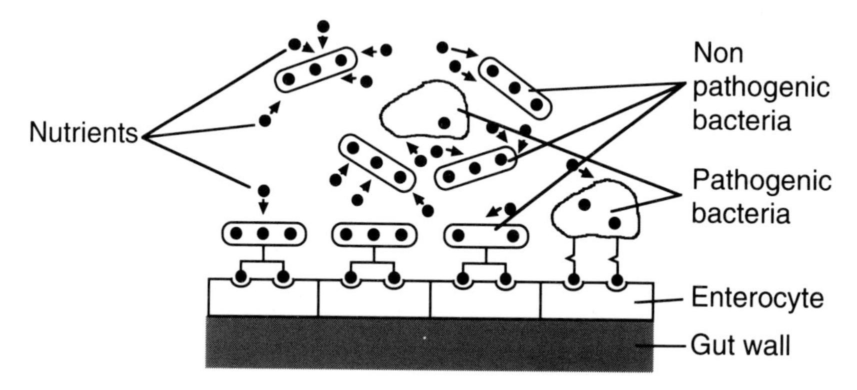

Figure 7.9 Non-pathogenic bacteria and pathogenic bacteria often compete for nutrients, e.g. carbon, nitrogen and minerals. Non-pathogenic bacteria can compete successfully and so colonise the intestine to a greater extent.

Competitive exclusion could also be as a result of aggregation of non-pathogens to pathogens, preventing binding to attachment sites and leading to their removal from the gut (*Figure 7.10*).

Species specificity

It has been suggested that the protective flora may be host specific (Snoeyenbos, 1979) but there is some degree of uncertainty about this process as some trials have shown that gut micro flora from both chickens and turkeys were reciprocally protective. However, it has been claimed that the turkey was less well protected than the chicken by competitive exclusion (Weinack *et al.*, 1982). Differences in specificity between these species may be present, with Mead and Impey (1984) finding no protection of turkeys from a partially defined flora that protected chickens.

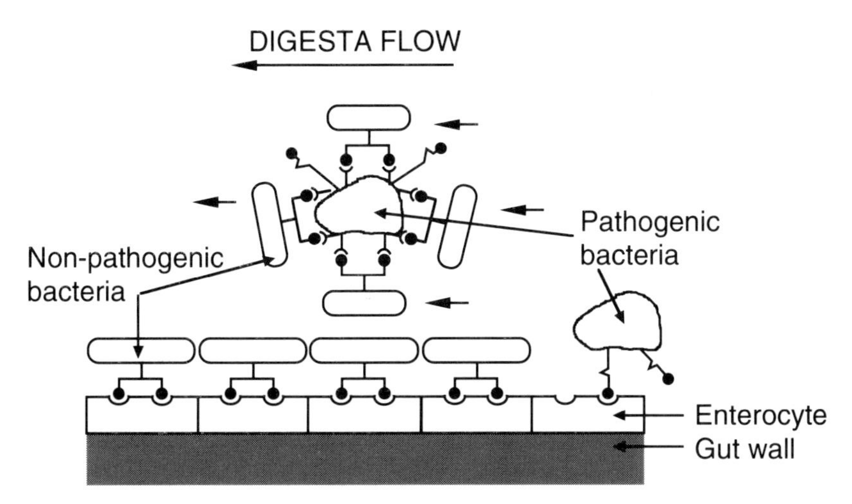

Figure 7.10 Accumulation or grouping of non-pathogens to pathogens in areas of quick digesta flow may allow their removal from the gut.

Reduction in toxic amine production

Metabolic activity of the intestinal micro-flora produces amines and ammonia which may have deleterious effects on the host animal. For example, amines produced after weaning have been found to be associated with diarrhoea (Porter and Kenworthy, 1969). They are irritating and toxic, and increase intestinal peristalsis, which may account for the diarrhoea (Muralidhara *et al.*, 1977). It has been shown that the level of amines produced within the gut can be reduced by *Lactobacilli* e.g. *L. acidophilus* (Schaedler and Dubos, 1962; Porter and Kenworthy, 1969; Hill *et al.*, 1970) and this may be important in maintaining a high health status. It is also claimed that they can detoxify pathogenic toxins, for example *L. bulgaricus* was shown to neutralize *E. coli* enterotoxin (Mitchell and Kenworthy, 1976).

Interactions with bile

Bile salts, produced within the liver, are surface-active chemicals which aid digestion by forming polymolecular aggregates with water-insoluble lipids and fat-soluble vitamins (Sandine, 1979). Bile is important in blocking the passage of many live organisms into the lower intestine and has been shown to inhibit the growth of enteric anaerobic bacteria, the mechanism of inhibition being described by many authors, eg. Floch *et al.* (1970). Specific strains of *Lactobacilli* can release free bile acids into the intestinal tract, and could, as a consequence of this, influence the balance of bacteria present within the gut (Sandine, 1979). Unconjugated (free) bile acids are much more inhibitory than conjugated forms (Floch *et al.*, 1972).

Antibiotic production

There have been several reports of antibiotic production by lactic acid producing species of *Streptococcus* and *Lactobacillus* (Whitehead, 1933; Mattick and Hirsch, 1944; Oxford,

1944; Su, 1948; Hirsch and Grinsted, 1951; Hirsch and Wheater, 1951; Wheater, Hirsch and Mattick, 1951; Vincent *et al.*, 1955; Vincent *et al.*, 1959). In a few instances (Whitehead, 1933; Oxford, 1944; Berridge, 1949), it was suggested that these antibiotics may have been polypeptides. Numerous antibacterial substances have been identified as products of lactic acid bacteria (Table 7.4).

Table 7.4 *Examples of antibacterial substances produced by lactic acid bacteria.*

Antibacterial Substance	*Source*
Nisin and diplococcin	Matlick and Hirsh (1944); Oxford (1944)
Acidophilin	Vakil and Shahani (1965)
Lactocidin	Grossowics (1947); Vincent et al. (1959)
	Shahani and Ayebo (1980)
Acidolin	Hamdan and Mikolajcik (1973)
Lactolin	Kodomar (1952)
Bulgarican	Reddy and Shahani (1965), Shahani and Ayebo (1980)

When tested, purified cell-free extracts have been shown to inhibit the growth of *Salmonella*, *Shigella*, *Staphylococcus*, *Proteus*, *Pseudomonas* and *E. coli*. However, little appears to be known about the structure and activity *in vivo* of these antibacterial substances (Shahani *et al.*, 1976; Tagg *et al.*, 1976; Sandine, 1979) and only specific strains of *Lactobacilli* have the ability to produce them. As well as lactic acid, other metabolites produced by *Lactobacilli* are known to inhibit bacterial growth, such as acetic acid which lowers the oxidation reduction potential and increases digestive movement (Savage *et al.*, 1969).

Lactocidin has been shown to be selectively active against parasitic and Gram-negative micro-organisms but not Gram-positive micro-organisms. However, it should be noted that Fuller (1989) claimed that these anti-bacterials detected *in vitro* may not be active in the intestine.

Bacteriocins are anti-microbial proteins which act specifically on target bacterial cells. They are self-destructive in that they lead to the death of the producer cell (Tagg *et al.*, 1976). They are usually water soluble and active at low concentrations (Hurst, 1981).

Production of organic acids

As already stated, lactic acid alone has been shown *in vitro* and *in vivo* in the pig, to inhibit the growth of Gram-negative pathogens such as *E. coli* (Tramer, 1966; and Cole *et al.*, 1968; respectively). This inhibition could, at least in part, be due to the effect of its pH. *E. coli* grows well at pH 8, so a lower pH will be inhibitory, with pH 4.5 bacteriostatic (Fuller and Brooker, 1974). Native *Lactobacilli* could therefore be producing lactic acid *in vivo* with the above effects.

Fermentation products vary qualitatively and quantitatively with the site of starch degradation and the type of carbohydrate in the diet. Lactic acid seems to be the principal end-product in stomach contents, while volatile fatty acids predominate in the caecal contents. Lactic acid bacteria, in particular certain *Lactobacilli*, are known to suppress the growth *in vitro* of other bacteria and some fungi (Savage *et al.*, 1968). Table 7.5 details some of the bacteria reportedly inhibited by *Lactobacilli*.

Table 7.5 *Antagonisisms exerted by* Lactobacilli.

Bacteria inhibited by Lactobacilli	*Authors*
Bacillus subtilis	Shahani, Vakil and Kilara (1976)
Serratia marcescens	
Proteus vulgaris	
P.fluorescens	
P.aeruginosa	
E. coli	
Sarcina lutea	
Streptococcus lactis	
Staphylococcus aureus	
S.aureus	Gilliland and Speck (1977)
S.typhimurium	
Enteropathic *E. coli*	
Clostridiune perfringens	

The presence of anti-enterotoxic activity

Toxins produced by pathogens at times of disease can bind to epithelial receptors preventing colonisation of bacteria (*Figure 7.11*). The enterotoxin, produced by a number of *E. coli* strains, that causes fluids to be lost from the intestine, can be neutralised by live probiotic type bacteria. This is the case with lactic acid bacteria (Mitchell and Kenworthy, 1976; Fuller and Cole, 1988). There is also anti-enterotoxic activity in *L. bulgaricus* and *S. faecium*, although it is not common (Mitchell and Kenworthy, 1976).

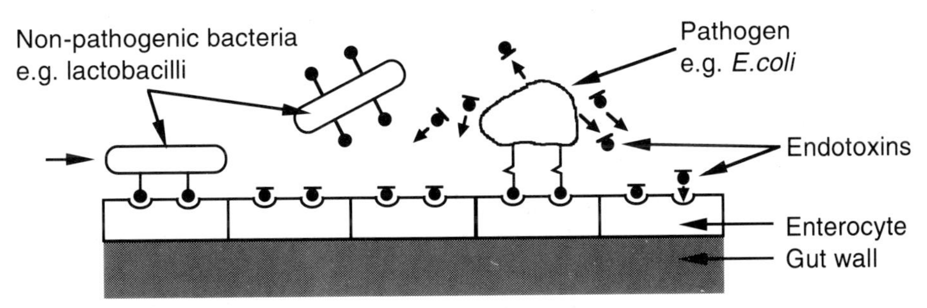

Figure 7.11 Endotoxins produced by pathogens at times of disease prevent colonization by bacteria.

Stimulation of immunity.

Immunity may be defined as the ability of the host to resist infection by bacteria and/or the harmful effects of their toxins. Thus, immunity may be natural (innate) or acquired.

Lactic acid bacteria may stimulate the production of antibodies and phagocytic activity against pathogens in the intestine and other tissues of the body (Fuller, 1989). It has been claimed that bacteria contain some common antigens (*Figure 7.12*) that can cross react with some pathogenic micro-organisms and thus immunize against an invasion of pathogenic micro-organisms (Perdigon *et al.*, 1986).

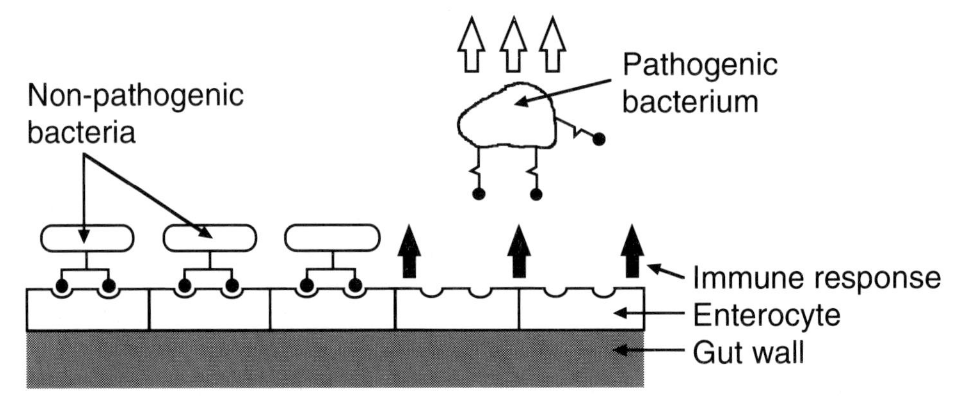

Figure 7.12 Antigens from non-pathogenic bacteria potentiate the hosts immune response to pathogens. Pathogens are repelled from enterocyte receptors.

Supplementary bacteria may also activate the immune system involved in resistance by their antigens, (Morland and Midtuedt, 1984; Perdigon *et al.*, 1986). *Lactobacillus* antigens could include lipoteichoic acids (Setoyama, *et al.*, 1985; and Op Den Camp *et al.*, 1985) and cell wall proteins (Conway and Kjelleferg, 1989). Increased leukocytic activity in gnotobiotic animals inoculated with *Lactobacilli* has been observed, which suggests that *Lactobacilli* may be involved in the immune response of conventional animals (Pollman *et al.*, 1980).

The role of bacteria in digestion of protein

In monogastric animals, microbial fermentation takes place mainly, but not exclusively, in the large intestine. The extent to which absorption takes place in the large intestine is open to some debate. Once ammonia has been absorbed, it cannot be used and is excreted (Just, 1983). Although a simple approach of assuming that digestion is by hydrolysis before the ileo-caecal junction and by fermentation in the large intestine is often used, this is not the case. For example, up to 20% of fermentation in pigs can take place ahead of the ileum with the advantage of enhanced absorption.

Bacteria are involved in the digestion of endogenous compounds. Where no bacteria are present, e.g. germ free animals, it has been shown that endogenous protein can collect in the caecum causing expansion of this organ (Luckey, 1963). Bacteria can also break down compounds produced by the gastro-intestinal tract but which are not hydrolysed by enzymes in the stomach and small intestine. These products can include urea (Levensen *et al.*,1959; Delluva *et al.*, 1968; Okumura *et al.*, 1976), bacterial mucoproteins, mucosal residues, mucus and uric acid (Mason, 1980). Bacteria inhabiting the caecum have also been shown to be capable of catabolising all L-amino acids *in vitro* (Fauconneau and Michel, 1970).

Breakdown of protein leads mainly to the production of ammonia but small quantities of amide, indole and phenolic compounds, keto acids, and carbon dioxide have been shown to be produced in the pig (Just, 1983). A beneficial side-effect is that bacterial action helps circulation of urea which may in turn help the production of non-essential amino acids.

Increased absorption and enzyme activity

The gastro-intestinal micro-flora is well known to influence the levels of absorptive enzymes in the microvillus membranes (brush borders) of intestinal epithelial cells in laboratory rodents. Although the mechanism by which it achieves this is unknown, it has been shown that gastric *Lactobacilli* can alter the levels of enzymes in germ free mice. This suggests that *Lactobacilli* may affect the absorption of nutrients by the animals, as the microvillus enzymes are known to mediate such absorption (Savage *et al.*, 1983). Bacteria are more likely to be involved in nitrogen metabolism when an amino acid or protein is in short supply (March, 1979).

Additional dietary *Lactobacilli* have been claimed to affect the metabolism of the hosts, micro-flora (Gilliland and Speck, 1977; Fuller and Cole, 1988), due to the effect of certain *Lactobacilli* on enzyme activity. It has been shown that *Lactobacilli* can increase or decrease enzyme activity, providing beneficial effects to the host (Reddy and Wastmann, 1966). Galactosidase is an enzyme of bacterial origin which is required by mammals to break down lactose (milk sugar). The feeding of yoghurt (containing *Lactobacillus bulgaricus* and *Streptococcus thermophilus*), has been shown to increase the concentration of this enzyme in the small intestine (Fuller, 1989).

Production of hydrogen peroxide

Under certain circumstances, some lactic acid-producing bacteria form detectable amounts of hydrogen peroxide (Sandine, 1979; Price and Lee, 1970; Speck *et al.*, 1970). This will inhibit the growth of many bacteria, especially pathogenic Gram-negative types. It can also be involved in the activation of the lactoperoxidase-thiocyanate system in the gut. In this system, lactoperoxidase combines with hydrogen peroxide and then oxidises thiocyanate to an intermediary oxidation product. This substance can inhibit bacterial growth, and may be bacteriocidal at a low pH.

Other effects of Lactobacilli

Lactobacilli and other non-pathogenic bacteria have also been claimed to be involved in:

- Vitamin production (Coates *et al.*, 1968; Henderickx *et al.*, 1980; Shahani and Ayebo, 1980; Carlstedt *et al.*, 1987).

- Lowering gut pH (Torrey and Kahn, 1923) and improving gastric homeostasis.
- The production of D-leucine (a by-product with inhibitory properties) (Gilliland and Speck, 1968).
- Production of other beneficial unidentified by-products (Hamdan and Mikolajcik, 1974).
- Deoxy-D-glucose production which has inhibitory properties (Muralidhara, 1974).
- Production of natural flavours and aromas to enhance food palatability (Schindler and Schmid, 1982) and possibly appetite stimulation.
- Production of beneficial nucleotides (Eyssen *et al.*, 1965).
- Changing the absorptive surface area of the small intestine (Rupin *et al.*, 1980; Savage, 1985; Savage and Whitt, 1982; Savage *et al.*, 1981; Umesoki *et al.*, 1982; Umesaki *et al.*, 1982; Whitt and Savage, 1981; 1988).
- Anti-microbial effects (McCormick and Savage, 1983).
- Anti-cholesteremic effects (Gilliland *et al.*, 1985).
- Anti-tumour effects (Friend and Shahani, 1984).

The responses to lactic acid producing bacteria sought in man are likely to relate to better health of the digestive tract while in domestic livestock, improved productive performance should also result. There is more quantitative evidence in the case of the latter due to the nature of the experimentation which is conducted. However, the use of fermented milk products by man is an age-old custom. *Figure 7.13* summarises the main effects of non-pathogenic bacteria with the gut of farm animals.

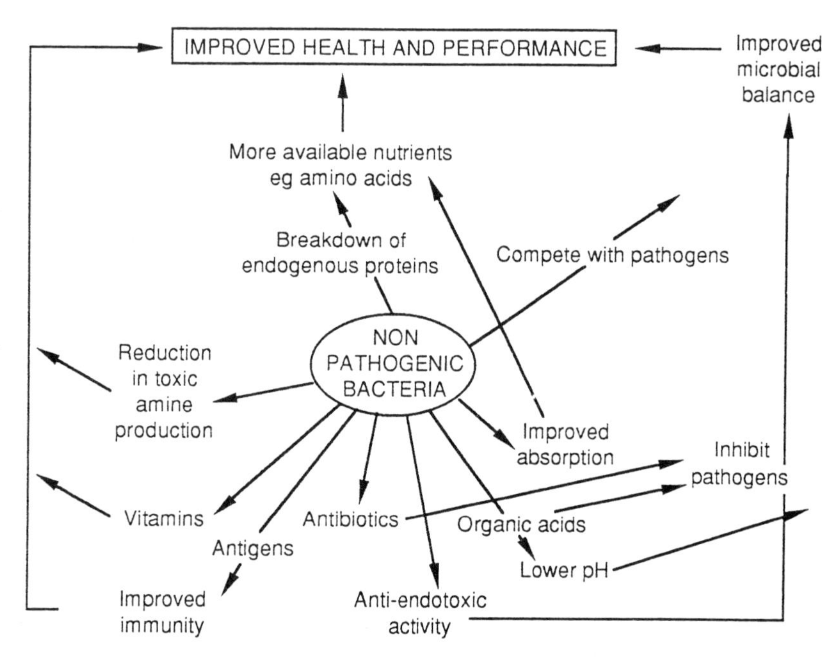

Figure 7.13 The effects of non-pathogenic bacteria in the gastro-intestinal tract of farm animals.

CHAPTER 8

Practical applications of micro-organisms in nutrition

Introduction

Manipulation of the diet has long been of interest in order to benefit the health and well-being of the animal. In both man and animals the health of the digestive tract is of importance. In farm animals the efficient conversion of food to product is a major target. In man, a quite opposite effect may be sought, where the desire is to eat a lot of food without much gain in weight. Many factors affect the chemical and biological processes in the gastro-intestinal tract. Efforts to modify and manipulate these processes have been developed commercially, particularly in the animal feed industry.

The composition of the diet, in terms of its nutritional value, can influence the working of the digestive system. Ingredients may, in addition to direct effects, have associative effects on other foods. For example, in monogastric animals the fibrous cell walls of cereals, which are used as various types of offals, can reduce the feeding value of the diet. They can reduce protein digestibility and may have their effects by physically enclosing valuable nutrients and also by increasing the rate of passage of digesta, giving less opportunity for degradation.

The nature of the diet can also alter fermentation patterns and this is particularly important in the ruminant animal. For example, roughage diets, high in cellulose, result in predominantly acetic acid as the end product of rumen fermentation. With largely concentrate diets, the proportion of propionic acid increases. The consequences of such changes are seen, for example, in the quality of milk as determined by its chemical composition.

It is the role of the nutritionist to take account of factors such as these, while at the same time balancing the diet to meet the requirements of the animal. In the monogastric animal particularly, this results in the use of high quality feedstuffs often competing with the needs of man. Consequently, there is interest in upgrading poor quality materials.

For example, fibrous plant materials can be subjected to physical treatment such as grinding, extrusion and pelleting.

On a world basis the potential of straw as a ruminant feed is considerable. The major drawback is low digestibility due to lignin, part of which comprises hemicelluloses. In order to break the bonds of the ligno-hemicellulose, treatment with sodium hydroxide has been used. There is also interest in further breakdown involving organisms which include soft-rot fungi, e.g. *Phanerochaete chryosporium*, and some bacteria, e.g. *Pseudomonas* species (Wallace *et al.*, 1983).

Nutrient availability

Enzyme treatment of materials to enhance hydrolysis before feeding is an attractive proposition. For example, lipases are widely used in the food industry. Packing house and abattoir wastes present particular disposal problems. If the materials involved can be upgraded to a useful animal feed, this represents a considerable benefit. Blood, bones and meat are typical materials that can have important uses.

If blood can be collected hygienically, it can be used to produce plasma protein which is valuable in the animal feed industry. However, the red cell fraction, which accounts for about 70% of the blood, presents a challenge, particularly relating to its colour. Typically, this can be treated after separation from the plasma by the addition of water and an alkaline protease at the rate of 0.4–0.5% by weight of blood. The red cells are completely broken down after four hours at 50°C and pH8. The process is stopped by inactivating the enzyme by reducing the pH to 4. The broth has its colour removed and is concentrated by filtration and evaporation respectively. The 40% solids material is then either spray or drum dried and, being 95% digestible, is suitable for incorporation into young animal diets. Similar techniques can be used to recover protein from bones and a neutral protease is used to make fish hydrolysate.

One of the most difficult animal waste products is feathers but considerable progress has been made in converting them to a useful feedstuff (Harvey, 1992). A major problem with feathers is that they contain keratin which, although high in cystine and cysteine, has a disulphide bond which is difficult to break. If it is broken by high temperature, there is a danger of damaging the protein. A new process (*Figure 8.1*) initially grinds the fresh feathers with a reagent for one hour at 60°C using a paddle mixer to break the disulphide bonds.

An enzyme cocktail, consisting of neutral protease and lipase, together with an emulsifier, is added. The feathers soon become reduced to a soup-like consistency, having about 20% solids. The enzymes are inactivated by reducing pH and the resultant material has potential value as a liquid feed. However, the greatest benefits are likely to be obtained in dry feeding systems where the slurry can first be concentrated by evaporation or by adding ground maize or soya bean meal before feeding. A practical application of this is to extrude the soya/feather blend and mix it with wheat and soya bean meal. It can be used for various purposes by varying the ratio of the materials. A typical analysis of such a material is given in Table 8.1 by Harvey (1992), who has reported successful feeding to poultry, pigs, fish and pets.

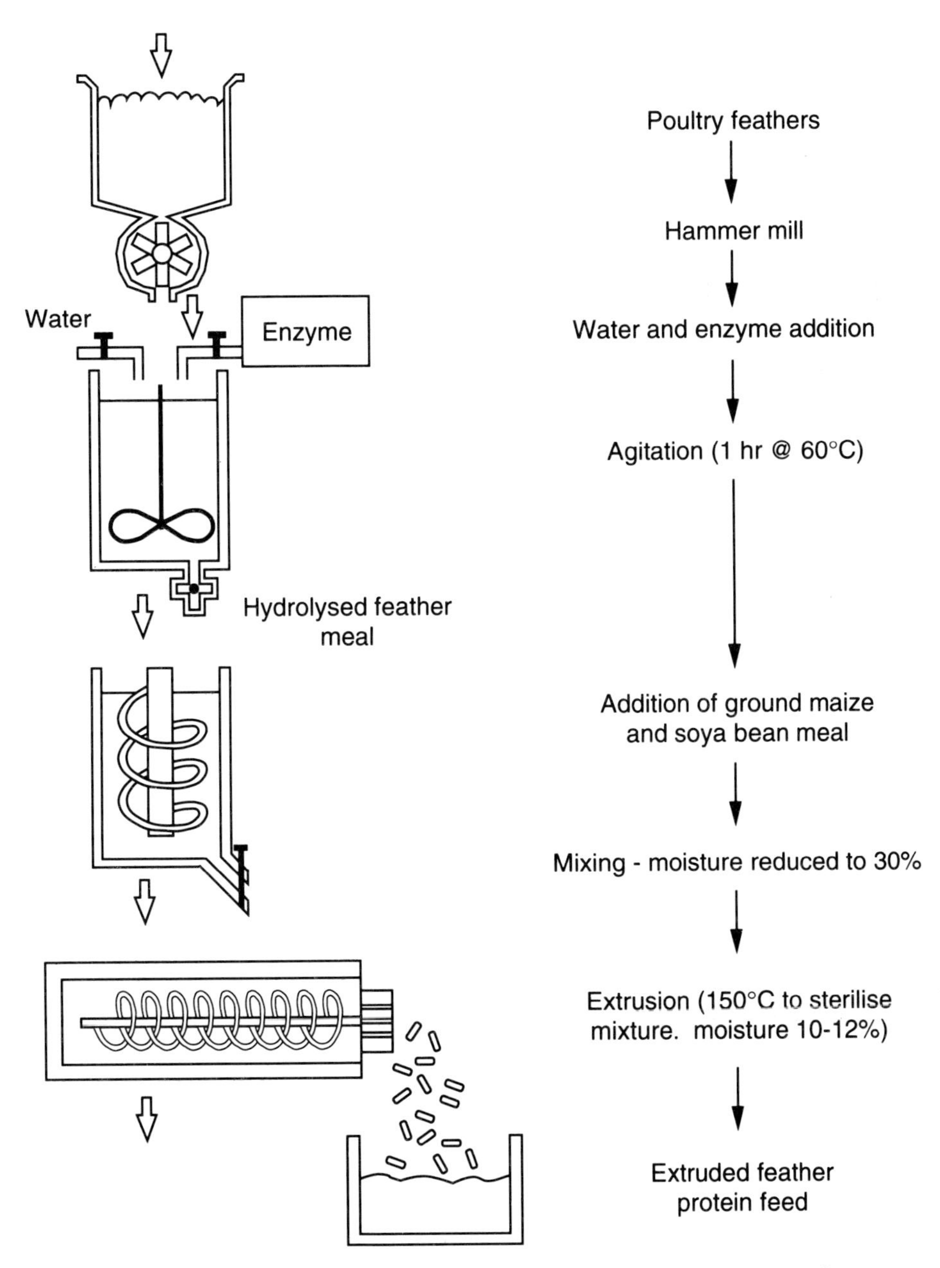

Figure 8.1 The processing of feathers with physical and enzymic treatment to produce a high quality animal feedstuff.

Table 8.1 *A typical feather meal product.*

Dry matter (%)	88.00
Fat (%)	1.50
Poultry ME (MJ/kg)	12.76
Crude protein (%)	45.00
Lysine (%)	1.35
Methionine and cystine (%)	3.09
Threonine (%)	2.10
Tryptophan (%)	0.38
Calcium (%)	0.13
Phosphorus (%)	0.54

Enzymes and the digestive tract

Enzymes are used widely in animal feeds, by mixing in prior to feeding in order to enhance digestion in the animal itself (Table 8.2). A number of such exogenous enzymes are commercially available and used in animal diets (Table 8.3).

The efficient output of quality end-products is no longer the sole target of animal production. In recent years there has been considerable emphasis on the reduction of pollution and care of the environment. Drastic remedies have been sought in some countries, for example Singapore, where pigs are banned. In others, for example the Netherlands, legislation places limits on the quantities of materials such as nitrogen and phosphorus that can be applied to the land (Table 8.4). In the latter case, great efforts are being made to control output by livestock. In the first instance this involves the use of much greater precision in the construction of diets, so that high quality diets offering optimum nutrient intakes result in lowered outputs. However, the use of enzymes in

Table 8.2 *Typically available commercial enzymes*

Enzyme	*Action*	*Typical Use*
Beta-glucanase	Beta-glucans to oligosaccarides and glucose	Barley based diets
Pentosanase	Pentosans to low molecular weight products and glucose	Wheat based diets
Amylase	Starch to dextrins and sugars	High fibre diets
Lipase	Fats to fatty acids	High fat diets
Proteinases	Protein to peptides	Numerous protein materials, e.g. soya and soya products
Phytase	Increased utilisation of phytate phosphorus	Phosphorus supplied by plant sources.

Table 8.3 *A typical commercial enzyme list together with the appropriate International Union of Biochemistry (IUB) numbers.*

Enzyme	*Product Name*	*Description*	IUB *Number*
Alpha-amylases	Fungal amylase (dedusted also available)	Fungal alpha-amylase (A.*oryzae* var.) for dextrinizing and saccharifying starch	3.2.1.1
	Bacterial amylase	Bacterial alpha-amylase (B.*subtilis* var.) for starch liquefaction at temperatures up to 90°C	3.2.1.1
	High temperature bacterial amylases	Thermostable bacterial alpha-amylase (B.*licheniformis* var.) for starch liquefaction at temperatures above 90°C	
Beta-glucanase	Bacterial beta-glucanase	Thermostable bacterial beta-glucanase (B.*subtilis* var.) for the hydrolysis of cereal beta-glucan polysaccharides	3.2.1.6
Catalase	Micro-catalase	A standardised liquid enzyme obtained by the controlled fermentation of *Micrococcus lysodeikticus* which catalyses the decomposition of hydrogen peroxide to water and molecular oxygen	1.11.1.6
Cellulases	Cellulase	Fungal cellulase system (A.*niger* var.) which is primarily active on soluble forms of cellulose.	3.2.1.4
	Cellulase TR	Multi-enzyme system (T.*reesei* var.) with endo- and exoglucanase activity	3.2.1.4
Glactomannase	Hemicellulase	Fungal hemicellulase (A.*niger* var.) specific for the hydrolysis of glactomannan gums and soluble cellulose	
Glucoamylase	Fungal glucoamylase	Fungal glucoamylase (A.*niger* var.) which is capable of hydrolysing both the linear and branched glucosidic linkages of starch and oligosaccharides resulting in essentially quantitative yields of glucose	
Lipases	Pancreatic lipase	Pancreatic derived lipase which hydrolyzes insoluble fats and fatty acid esters to yield monoglycerides diglycerides, glycerol and liberation of free fatty aicds	3.1.1.3

Table 8.3 *Continued.*

Enzyme	*Product Name*	*Description*	IUB *Number*
Lactase	Fungal lactase	Fungal lactase (*A.oryzae* var.) which hydrolyses lactose, forming glucose and galactose	3.2.1.108
Pectinase	Pectinase XL	Concentrated fungal pectic enzyme system (*A.niger* var.) for efficient depolymerisation of naturally occurring pectins	3.2.1.15
	Pectinase AT	Fungal pectic enzyme system (*A.niger* var.) especially effective in low pH processes such as cranberry juice production	
	Pectinase ML	Fungal pectic enzyme system (*A.niger* var. and *T.reesei* var.) used in fruit maceration or liquefaction processes to maximise fruit juice and solids extraction.	
	Pectinase APXL	Fungal pectic enzyme system (*A.niger* var.) used for hydrolysis of both soluble and colloidal pectin substances. Also high in arabinase activity for elimination of araban haze.	
Proteases	Bacterial protease	Bacterial proteases (*B. subtilis* var.) which effectively hydrolyse proteins over the neutral to alkaline pH range	3.4.24.28
	Acid fungal protease	Acid fungal protease (*A.niger* var.) characterised by its ability to hydrolyse proteins under acidic conditions. (pH 2.5–3.5)	
	Fungal protease (dedusted also available)	Fungal protease (*A. oryzae* var.) containing endo- and exo- peptides with a broad substrate specificity for catalysing the hydrolysis of proteins	3.4.23.18
	Bromelain	A protease isolated from the pineapple plant which hydrolyzes plant and animal proteins to peptides and amino acids	3.4.22.33

Table 8.3 *Continued.*

Enzyme	*Product Name*	*Description*	IUB *Number*
	Papain	A protease isolated from the papaya latex. The enzyme extensively hydrolyses proteins and has excellent stability at elevated temperatures	3.4.22.2
	Papain 6,000	A liquid protease used for hydrolysing proteins	
	Alkaline protease	Bacterial alkaline proteases, (*B.licheniformis* var.) for hydrolysing proteins under highly alkaline conditions	

(*Kindly provided by Valley Research Inc., South Bend, IN, USA*)
The IUB (1991) number relates to the specificity of the enzyme. Each number from left to right defines the enzyme more specifically. For example, alpha-amylase 3.2.1.1 is translated as follows:

3–the first number shows to which of the six main divisions (classes) the enzyme belongs
2–the second figure indicates the subclass
1–the third figure gives the sub-subclass
1– the fourth figure is the serial number of the enzyme in its sub-subclass

the diet will be an important factor in minimising pollution. The needs of efficient production and care of the environment are entirely compatible.

It is important when trying to improve the digestibility of the diet that the correct enzymes are chosen for the correct substrates. Consequently enzyme mixtures should be chosen for each particular diet and it is necessary to consider the response of the various nutrients individually.

Phosphorus

The problem of excessive phosphorus excretion is well established and serves as a good example of the nutritionists' attention to the use of exogenous enzymes. Phosphorus has many roles in the body with the result that it is required in large quantities by the animal. It plays an important part in the formation of bones and is involved in muscle and nerve metabolism, energy transformation and acid-base balance of body fluids.

Table 8.4 ***Permissible levels of phosphate (kg P_2O_5/ha) on farm land in the Netherlands.***

	1990	*Up to 1994*	*After 1994*
Arable	125	125	125
Grass	250	200	175
Maize	350	250	175

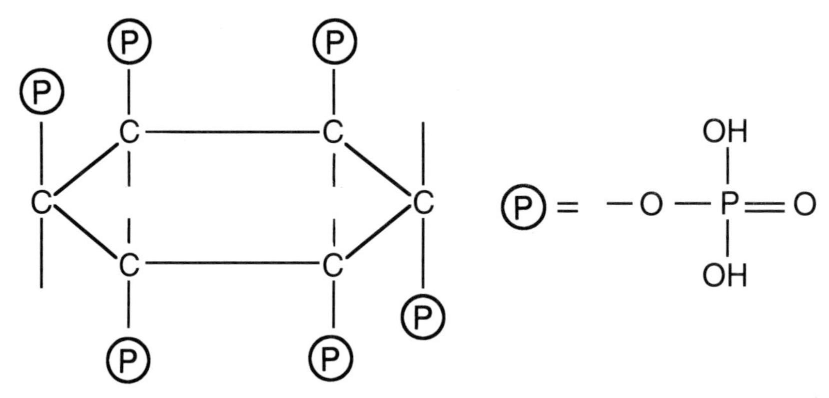

Figure 8.2 Phytic acid. A large amount of the phosphorus in plants is in the form of phytic acid salts, called phytates.

Two important factors are involved in minimising excretory phosphorus from animals, namely, its bio-availability in various feedstuffs and the level of supply. While phosphorus is from plant, animal and inorganic sources, it is the plant ingredients that have the major problem of low availability. This results from phosphorus being in the form of phytate (or phytin), a salt of phytic acid (*Figure* 8.2). Phytases in plant material have generally been attributed little value. However, Jongbloed and Kemme (1990) demonstrated that wheat phytase, for example, could improve digestibility of phosphorus, by pigs, from 27 to 50%.

Many micro-organisms produce phytase and, commercially, members of *Aspergillus spp*. have proved successful. For example, the use of a culture of *Aspergillus ficuum* in the diet of poultry was shown to give a considerable increase in bone ash content (Nelson *et al.*, 1968). More recently, Simons *et al.* (1990) have shown an increase in digestibility of phosphorus from 20 to 46% with the addition of microbial phytase to maize/soya diets for pigs. It has been reported that this greater breakdown of phytic acid took place mainly in the stomach (Jongbloed *et al.*, 1990). It has been shown in young pigs (10–30kg liveweight) that feed intake and growth rate were increased by the addition of microbial phytase, as well as improving digestibility of phosphorus by 20% (Beers and

Table 8.5 *A prediction of the reduction in phosphorus excretion in pigs in the Netherlands.*

Improvement Factor	*Intake (kg/animal)*	*Excretion (kg/animal)*	*Excretion (% 1983 value)*
1983	1.61	1.18	100
1990	1.23	0.83	70
Using microbial phytase	1.01	0.61	52
Using alternative feedstuffs	0.89	0.49	42

(Coppoolse *et al.*, 1990)

Koorn, 1990). Even higher responses are achieved with poultry. Phytase is commonly used at about 400–500 units of activity/kg feed.

Predictions from the Netherlands suggest that large reductions in phosphorus excretion are possible through a better knowledge of requirements, a judicious selection of feed ingredients and the use of microbial phytases (Table 8.5).

Other dietary enzymes

Dietary enzymes are used in situations when it is considered that the animal is unable to secrete sufficient of its own appropriate enzymes. They can be used in a variety of circumstances, for example, in the young animal with an immature digestive enzyme system. However, the fibrous cell walls of many plant feedstuffs are a particular problem for non-ruminant animals.

The major factors causing problems are the non-starch polysaccharides (NSP) which reduce the feeding value of many plant materials in the diet of monogastric animals. These animals have an inadequate production of enzymes, which affect hydrolysis and absorption. They do, of course, subject them to fermentation in the later stages of the gastro-intestinal tract, but absorption is often poor at this point.

NSP vary from simple polysaccharides, such as β-glucans, to more complex arabinoxylans (pentosans). The important NSP in cereals are pentosans and β-glucans while in legumes they tend to be pectic substances (which are rich in uronic acid) and flatulence producing oligosaccharides (Table 8.6). The flatulence producing

Table 8.6 *Anti-nutritional factor content of some common feedstuff (g/100g DM).*

			Flatulence producing oligosaccharides		
Feedstuff	**β-glucan*	*#Pentosan*	*Raffinose*	*Stachyose*	*Verbascose*
Barley	4.3	6.0	0.2		
Oats	3.4	6.6		0.5	0.3
Rye	1.9	9.3	0.8		
Wheat	0.7	6.6	0.7		
Triticale	0.7	7.1			
Sorghum	1.0				
Beans (*Vicia faba*)			0.2	0.7	2.0
Beans (*Phaseolus vulgaris*)			0.5	2.6	0.1
Beans (*Glycine max.*)			2.3	1.9	0.1
Peas			0.7	2.3	2.1
Sunflower meal			1.9	0.4	
Cottonseed meal	3.6	0.9			
Soya bean meal			0.7	4.1	
Rape seed meal			0.5	1.4	

*(1–3), (1–4) β-glucans (30:70) mainly soluble.
Arabinoxylan (60 xylose:40 arabinose)
(Dierick, 1994)

oligosaccharides (raffinose, stachyose, verbascose and ajucose) are resistant to digestion and pass to the hind gut where they are fermented and can produce large amounts of gas.

In farm livestock particular emphasis has been given to improving the feeding value of cereals. β-glucans are a problem in barley and oats, with pentosans being of particular significance in wheat, triticale and rye. Perhaps the two major characteristics of β-glucan which affect barley are reduced feeding value and increased viscosity of the excreta which can be a big problem with poultry (e.g. sticky droppings). There is a good relationship between intestinal viscosity and reduced feeding value in both pigs and poultry. The use of β-glucanase to improve the feeding value of barley is well established. In general, it is used to bring the feeding value of barley up to that of other cereals (e.g. Pugh, 1992 (Table 8.7); Skoufos and Fthenakis, 1992).

Table 8.7 *The use of β-glucanase to improve the feeding value of barley for broiler chickens.*

	Monocereal wheat	*25% barley + β-glucanase*
No. of birds	45,000	45,000
Age (days)	47	47
Liveweight (g)	2,073	2,082
Food conversion ratio	2.10	2.09
Mortality (%)	7.7	7.3

(Pugh) 1992)

The addition of pentosanase to improve the feeding value of, for example, rye for pigs has been well demonstrated. Improvements of 7% and 8% have been reported for growth rate and feed utilisation respectively (Thacker, 1988). Such differences have been associated with improvements in ileal digestibility of dry matter, nitrogen and NDF of 3.1, 4.2 and 7.2%, respectively (Buraczewska, 1988).

From a commercial point of view, wheat is a much more important feedstuff. Annison and Chocht (1994) have reported improvements in feeding value to poultry by adding xylanases. Work at ADAS Gleadthorpe Experimental Husbandry Farm in the United Kingdom has shown the value of pentosanase in wheat diets for broiler chicks (Tucker, unpublished), particularly up to 21 days. The latter observation reinforces the view that young animals (e.g. the chick and the weaner pig) are most responsive to enzyme supplementation.

Absorption enhancement

The need for absorption to follow any improvements in hydrolysis or fermentation is worthy of emphasis. The efficacy of this will depend, to some extent, on the site of digestion and on the opportunity for absorption in the various species. An example of

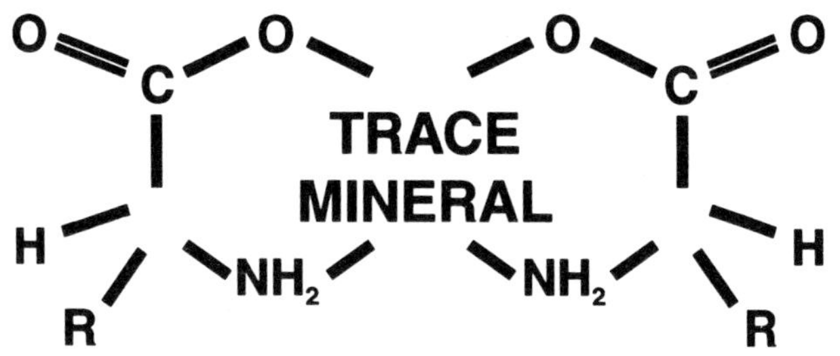

Figure 8.3 A trace mineral chelate (or proteinate) is a transition element, such as zinc, copper, manganese, iron or cobalt, that is chelated to, for example, amino acids and/or peptides. The chelate that results from the binding of the mineral and amino acid/peptide carries no electrical charge, and as such is stable through the pH changes that take place during digestion. As a result, instead of becoming insoluble, the mineral is now in a physiologically acceptable form to be absorbed into the bloodstream.

a specific effort being made to improve absorption is chelated minerals. Trace minerals (e.g. zinc and manganese) are generally added in the form of oxides or sulphides. These are ionised at the low pH of the stomach and not well absorbed. In this form, e.g. Zn^{++}, they are able to react with other products of digestion. However, for ease of absorption a complex with an easily absorbed organic compound (e.g., an amino acid or small peptide) should occur. This is a chelate, an individual mineral molecule chemically bonded to an organic ligand. The elements which are capable of producing chemical chelates are the transition elements (Mn, Fe, Co, Cu and Zn). An example of a chelate or proteinate is given in *Figure* 8.3. Generally, a mineral bonded to a single amino acid is considered to be a chelate and to a peptide a proteinate. Their effects have generally been in increasing the availability of the elements concerned and in alleviating "stress" conditions. Furthermore, particular organs can be targeted by the choice of the appropriate ligand. A classic example of this is the use of zinc methionine to improve the quality of the coat of hair in dogs and horses. Recently there is evidence of influence of chelated minerals on other organs and functions.

The gut flora

Manipulation of the gut micro-flora to influence the health and well-being of animals and man may be achieved by materials which are generally regarded as having probiotic effects. These are:

- Live organisms
- Organisms which have been killed
- Chemicals

In this context live *Lactobacillus* and *Streptococcus* have received particular attention because of their central role in lactic acid production and because of their commercial development. However, other micro-organisms also have important functions in modifying the processes of the digestive tract. Such modifying roles should not be considered only in terms of bacteria. For example, various fungi (e.g. *Saccharomyces* and *Aspergillus*) have been shown to have important effects, particularly where fermentation is taking place. Many groups of non-pathogenic micro-organisms have been extensively investigated.

Micro-organisms in probiosis

A major use of micro-organisms is as probiotics, which are often called direct-fed microbials. The current theory regarding the commercial use of probiotics is that the microbial balance can be tipped in favour of the beneficial non-pathogenic types by several methods.

The main method to be considered in this chapter is the addition of live beneficial micro-organisms. To date these have been predominantly *Lactobacilli* and *Streptococci*, although others are now available. In order to derive any benefit from their addition it is claimed essential that the viability of the preparation is maintained during storage and application to feeds, particularly if they undergo processing before feeding. Another approach is to use bacterial spores as these are more resistant to high temperature and high moisture conditions (for example, associated with pelleting), than live cultures. However, they must germinate in the small intestine if they are to be of benefit to the animal.

The characteristics which need to be fulfilled by an organism in order to be useful as a "live" probiotic/direct fed microbial include:

- **Not harmful to the animal** The organisms should not cause disease nor should they be toxic (Fuller, 1989).
- **Acid and bile resistant** If a live micro-organism is to survive passage through the stomach and reach the intestine alive it must be acid and bile resistant. However, it is likely that they may be protected by the ingested food. Micro-encapsulation, for example, using β-glucans, has been used as a means of increasing the ability of the organism to survive.
- **Ability to colonise the gut** Only specific bacterial strains are able to effectively adhere to the intestinal epithelium, which is necessary if their mode of action is competitive exclusion. For example, it has been suggested in the pig, that only bacteria isolated from domestic pigs and the closely related wild boar are able to adhere to the squamous epithelial cells of the stomach (Barrow *et al.*, 1980). Thus, it is important to select bacteria common to the gastro-intestinal tract of the host animal (Shahani and Ayebo, 1980; Gilliland, 1987; Jones and Thomas, 1987). They will need to possess the appropriate colonisation or attachment factors (Fuller, 1989).
- **Ability to inhibit pathogen activity** The probiotic should be able to show its inhibitive powers towards pathogenic bacteria in the laboratory, if live microbial

activity is required. The organisms chosen should produce acid or other materials so aiding in the inhibition of Gram-negative pathogenic micro-organisms, such as *E. coli* (Tramer, 1966; Sandine, 1979).

- **Stable and viable under manufacturing conditions** Organisms in "live" probiotic products must be able to withstand freezing and the high temperatures of processing to be a viable product.
- **Stable and viable under storage conditions** If an anaerobic bacterium is chosen, it must be kept in anaerobic conditions until it is fed to the animal, if it is to survive (unless it is a spore former). Strictly speaking, anaerobic bacteria cannot be used in feed if viability is to be guaranteed and is of importance. The micro-organisms would also need to be kept at very low temperatures to avoid sporing and/or death.

Bacteria that only remain viable for short periods of time are unsuitable for commercial application as live probiotics. Facultative organisms such as *L. acidophilus* are not harmed by the presence of oxygen and, thus, are often selected for this reason.

Dosage

The minimum effective dose of live bacteria is not easily identified (Fuller, 1989). In trials about lg/kg feed is commonly used. Commercial probiotic products come in various forms, such as pellets, powder, capsules, paste and granules. All of the product is not, therefore, active biomass, making it difficult to assess true dose level. A typical live probiotic called Lacto-Sacc was stated to consist of 100 million CFU's/g *L. acidophilus* and 70 million CFU's/g of *S. faecium* compared with populations of gut micro-flora supposedly in the region of 10^{14} micro-organisms. *Figure 8.4* illustrates the perceived benefits of feeding live non-pathogenic micro-organisms to the new-born animals.

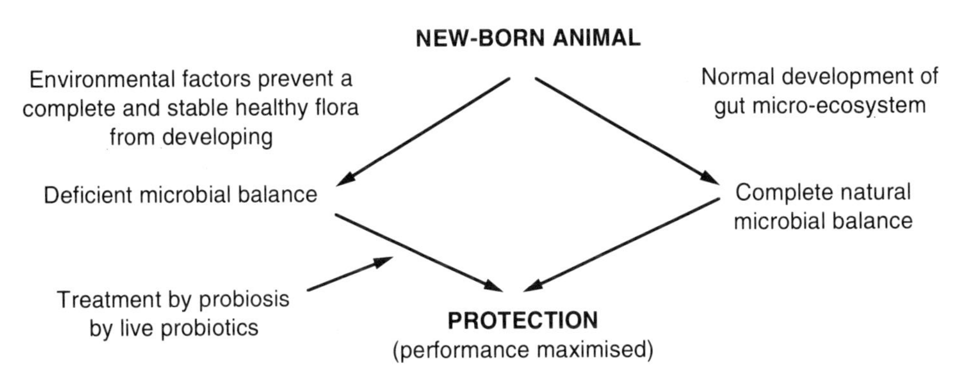

Figure 8.4 Theory of gut protection. When bacterial additives are fed to newborn animals with an incomplete natural flora, a stable population forms and the host animal is protected from the effects of pathogenic organisms.

Table 8.8 *Some common organisms used as direct-fed microbials.*

BACTERIA	*Lactobacillus*	*acidophilus*
		bifidus
		brevis
		bulgaricus
		casei
		cellobiose
		curvatus
		delbruecki
		fermentum
		lactis
		plantarum
		reuterii
		salvaricus
	Bacillus	*cereus*
		coagulans
		lentus
		licheniformis
		pumilus
		subtilis
		toyoi
	Bacteroides	*amylophilus*
		capillosis
		rumicola
		suis
	Bifidobacterium	*adolescentis*
		animalis
		bifidum
		infantus
		longum
		thermophilum
	Streptococcus	*cremorius*
		diacetilactis
		faecium
		intermedius
		lactis
		thermophilis
	Pediococcus	*acidilacticii*
		cerevisiae
		faecium
		lactis
		mesenteroides
		pentosaceus
		thermophilus
	Leuconostoc	
FUNGI	*Saccharomyces*	*cerevisiae*
		boulardii
	Totulopsi	*candida*
	Aspergillus	*niger*
		oryzae

Types of direct-fed microbial organisms

Many organisms have been used commercially to produce direct-fed microbials (Table 8.8). The most common are:

Lactobacilli

Lactobacilli are commonly used in probiotics as they are known to be non-pathogenic and are also natural inhabitants of the gastro-intestinal tract with many beneficial effects. It is also generally assumed that they are responsible for the health benefits associated with yoghurt.

Bacillus subtilis

This is a spore-forming organism and therefore more stable than *Lactobacilli* and *Streptococci*; viability should, in theory, be less of a problem. It has been claimed that this *Bacillus* species could reduce the number of *E. coli* in faeces, the digestive tract and blood (Pollman, 1986). Fuller (1989) has stated that *B. subtilis* is not an intestinal organism, but is a strict aerobe and he questioned its use as a live probiotic. However, it is established that pools of oxygen may be present in the gut of monogastric animals, e.g. the young pig (Stewart *et al.*, 1993).

Streptococci

Streptococci were originally classified as part of the Lactobacillaecae family, due to their common physiological traits. However, it is now a member of the Streptococcaecae family (Deibel and Seeley, 1974) and they are often called *Enterococci*. They are Gram positive, facultatively anaerobic, non-motile cocci, spherical or ovoid, occurring in pairs or chains. They have complex nutritional requirements and have a fermentative metabolism (homo-fermentative), producing D-lactic acid from glucose (Deibel and Seeley, 1974). They are located in the mouth, and in the entire intestinal tract. They are very tolerant of a range of temperatures, bile, NaCl and low pH (Frobisher *et al.*, 1974). *Streptococci* are lactic acid bacteria which have been claimed to have a role in maintaining a balanced micro-flora and for this reason have been selected for their probiotic potential. However, less attention has been given to *Streptococci* than *Lactobacilli*, probably because some species of *Streptococcus* can be pathogens. *Streptococci* are found attached to the squamous epithelium but in lower numbers than *Lactobacilli*. This is probably due to their greater sensitivity to pH and pepsin (Barrow *et al.*, 1980).

Fungi

Fungi too, have been used to manipulate the gut environment. As far as farm livestock are concerned this has been largely, but not exclusively in the ruminant. In this respect *Saccharomyces cervisiae* and *Aspergillus* have received most attention although there is current interest in the use of *Saccharomyces boulardii* in the diets of monogastric animals. The former have been used to alter microbial digestion in the rumen and so enhance productivity.

Responses to lactic acid producing bacteria

Animal Responses

In recent years, interest has grown in the feeding of micro-organisms, particularly *Lactobacilli*, as an alternative to the use of antibiotics and also following antibiotic therapy. Antibiotic therapy often lowers the *Lactobacillus* population in the intestinal tract and it is claimed that replenishing the intestinal tract with, for example, *L. acidophilus* results in an accelerated return to a beneficial intestinal population. Modern day farming often stops the young animal from obtaining its normal bacterial loading. It has been suggested that the most satisfactory method to overcome this is to feed bacteria. As stated before, the best results are claimed to be obtained from the ingestion of $1x10^8$ to $1x10^9$ viable *L. acidophilus* daily, but ingestion of excessive numbers may induce diarrhoea. Relative to humans, there is considerable information on farm animals, with further results appearing constantly. Improved gut health will result in more efficient function, e.g. digestibility (Table 8.9).

Table 8.9 *The influence of the inclusion in the diet of a mixture of* Lactobacillus acidophilus, Streptococcus faecium *and* Saccharomyces cerevisiae *in pig diets on ileal digestibility (%).*

	Control	*Treated*	*Change (%)*
Maize			
Protein	78.9	80.6	2.1
Lysine	84.2	89.9	6.8
Methionine	93.9	90.6	8.0
Threonine	77.5	75.3	10.1
Wheat			
Protein	80.7	85.2	5.6
Lysine	63.5	82.1	29.3
Methionine	72.7	83.1	14.3
Threonine	74.5	80.2	7.6
Soya bean meal (48% crude protein)			
Protein	77.2	85.0	10.1
Lysine	84.2	88.3	4.9
Methionine	75.1	88.4	17.7
Threonine	79.3	85.3	7.6

(Tossenberger J.Z. quoted by Gombos, 1991)

Times at which bacterial additives may be most effective

When considering the various classes of farm livestock, it is possible to highlight particular occasions when probiotics are most likely to have a positive effect. These were listed by Ewing and Haresign (1989) (see Table 8.10).

Table 8.10 *Times at which bacterial additives may be most effective.*

Calves	• After birth – to encourage the early establishment of beneficial micro-flora. • Change to bucket feeding • Before and after transportation • At weaning • Following over-eating or antibiotic administration
Adult cattle	To restore the desirable microbial balance and appetite following: • Ketosis • Antibiotic treatment for therapeutic reasons • Bloat • Difficult calvings
Lambs	• Weak or mismothered lambs should be treated as soon as possible after birth to encourage the early establishment of beneficial micro-flora • During change over to milk • If digestive upsets occur • Tailing and castration • Weaning onto concentrate feeds at young ages
Adult sheep	To restore the microbial balance and appetite following: • Twin lamb disease • Difficult lambings • Antibiotic treatment for therapeutic reasons
Piglets	• To prevent scouring in young piglets often associated with such stress as teeth clipping/iron injection, castration, weaning, and transportation • Following antibiotic treatment for therapeutic reasons
Older pigs	To restore the gut microbial balance and appetite following: • Antibiotic treatment for therapeutic reasons • Farrowing, particularly after a difficult farrowing • Transportation/mixing

(Ewing and Haresign, 1989)

Challenge

It has been suggested that, in the case of bacterial feed additives and probiotic like materials (and other additives generally), "response is dependent on challenge" (Cole, 1990). In other words, the greater the problem, the greater the response. For example, in a herd where weaner pigs had shown responses to organic acids in the drinking water, inconclusive responses were obtained when haemoyltic *E. coli* populations had reduced and diarrhoea problems diminished (Cole *et al.*, 1970).

In an attempt to rationalise the differences in response relating to the extent of the problem, data from a number of trials have been collected (*Figure* 8.5). Suckling piglets of sows fed probiotics based on lactic acid bacteria in the diet had lower mortality than

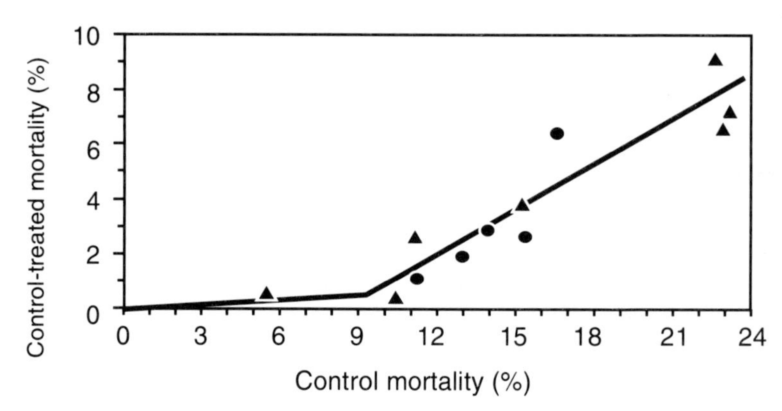

Figure 8.5 Piglet mortality (birth to weaning) when the sows were given probiotics in the diet. The degree of response to a probiotic is markedly influenced by the level of performance already existing in the herd. It is proposed that little or no response can be expected when pig mortality is below about 10% (Cole, 1991).
NB. Trials analysed all used *Lactobacillus* species. Each point represents an experiment. Slope of response above 10% mortality $Y = 0.553 \times X - 4.63$.

those which did not. However, the response was greatest when the control pigs had the highest mortality, i.e. the challenge was greatest. Presumably, below 10% piglet mortality, problems were not of an entirely enteric nature.

In this case, only the sow's food contained probiotic as the piglets were not offered supplementary or creep feed for their nutrients. The object was to make the sow's faeces less harmful by reducing the populations of harmful *E. coli*, etc., as suckling pigs are known to eat considerable quantities of faeces and bedding (see Table 4.3). Thus, it is postulated that the probiotics were having a beneficial effect on the micro-flora of the sow's digestive tract and consequently the faeces, which would cause less enteric problems to the piglets.

The young animal

Probiotics have been particularly effective in young animals, for example rabbits (Hollister *et al.*, 1991), chickens (Wiseman, 1990) calves (Rosell, 1987) and pigs (Cole, 1991). Presumably the developing digestive tract is vulnerable to microbial stimulation and any move to improved gut health is reflected in better general health, together with enhanced animal performance, e.g. particularly beneficial responses in terms of reduced mortality.

As far as the pig is concerned, perhaps one of the most essential times to maintain gut health is in the period immediately after weaning. This is well known as a period in which diarrhoea, growth checks and mortality can be a problem. Certainly one of the predisposing factors is the change from milk to a solid diet and its associated changes in the digestive system. At the same time, of course, there are dramatic changes in the enzyme system. In creating weaner diets, nutritionists generally aim to include some dairy product (e.g. dried whey). Such materials contain lactose and there is evidence that the combination of a probiotic based on lactic acid bacteria, together with a lactose source, presents a more powerful way of controlling haemlolytic *E. coli* than either given alone. Furthermore, a combination of lactic acid bacteria, organic acids and a source of lactose has been shown to be particularly effective in the control of haemolytic *E. coli*.

Adult animals

Adult animals which have a well balanced and stable gastro-intestinal micro-flora are less likely to be colonised by additional micro-organisms which enter the tract (Savage, 1977), as the adult animal's existing micro-flora is more likely to exclude the additional bacteria than *vice versa*. Consequently, there is less effect of probiotics in older animals than in young animals.

Laboratory inoculation of chicks has been shown to differ from commercial situations (Weinack *et al.*, 1981). The native gut micro-flora is more effective in influencing the demonstrable level of *E. coli* following experimental inoculation than in altering the levels of those *E. coli* naturally existing in the chick. Thus, 'normal' *E. coli* may be part of the natural microbial flora necessary for exclusion of other foreign bacteria. Probiosis within the gut will naturally try to avoid the colonisation of any foreign organisms, whether pathogenic or not. In the literature there have also been studies showing no correlation between gastro-intestinal and faecal bacterial populations (Paul and Hoskins, 1972; Muralidhara *et al.*, 1977; Pollman *et al.*, 1980) which would question the use of faecal analysis as a measure of gastro-intestinal bacterial change. Also, much of the *in vivo* research work has been carried out on germ-free animals. However, as an animals typical environment is far from germ-free this is not necessarily a relevant comparison.

Just as all antibiotics do not have the same mode of action it is likely that all probiotics do not either. Therefore, stating that probiotic results are variable may not reflect fairly the achievable response to all of the products, and this has already been discussed in relation to the suckling piglet.

Further reasons given to explain the variability in response to direct-fed "live" microbials are:

- Failure of the bacteria to survive storage, processing and gastric acids.
- Failure of the bacteria to implant in the gut.
- Destruction of the bacteria by antibiotics.
- Lack of stress in the animals and a *Lactobacillus* population in the gut which was not deficient, with the consequence that performance was already at a maximum.

Bacterial supplements are not only sold for inclusion in livestock feed but are also sold on the human health market. A comprehensive review of the health benefits from such supplements has been made by Sandine *et al.* (1972), and include the following:

- Reduction in *Candida* infections (Tomoda, 1983).
- Reduction in constipation (Alm *et al.*, 1983; Graf, 1983).
- Reduced lactose intolerance (Goodenough and Kleyn, 1976; Savaiano *et al.*, 1984).
- Stimulation of immunity (Perdigon *et al.*, 1986a,b).

- Reduction of serum cholesterol levels and consequently cardiovascular disease (Mann, 1977; Gilliland *et al.*, 1985; Nahaisi, 1986; Danielson *et al.*, 1989).
- Prevention of carcinogen production, e.g. nitrosamines (Nahaisi, 1986; Rowland and Grass, 1975).
- Protection against bowel cancer (Goldin and Gorbach, 1984).
- Reduction in gastro-intestinal pathogens (Gandhi and Nambrudripad, 1978).

Fungi

Yeast

The use of yeast can be considered in two ways. Firstly, it can, in sufficient quantities, act as a protein source, and brewers yeast has been used in this way for many years. Secondly, it can be used in small quantities, as live culture in the diet, to alter gut function. In this context, it has found a major role in the modification of rumen fermentation and consequently animal performance and also has effects in non-ruminants. In this context, it is not possible to use any yeast and considerable effort has been put into selecting the appropriate strains. *Saccharomyces cerevisiae* has received greatest attention but not all its strains can stimulate, for example, cellulose degradation in the rumen to the same extent (Table 8.11).

Table 8.11 *Influence of strain type of yeast on rate of cellulose digestion in cultures with* Fibrobacter succinogenes.

Source of Strain	*Strain number*	*Lag time (hours)*	*Rate of cellulose digestion (mg/h)*
Control (no yeast)	–	60.1	0.42
Yeast extract	–	60.7	0.52
Yeast	1026	42.7	0.38
American type culture collection No. 9763	430	45.1	0.28
American type culture collection No. 9080	432	48.6	0.50
Corn Silage	TG4	39.6	0.58

(Dawson, 1990)

Low levels (10–100 g/day) of fungal probiotics have proved effective at improving ruminant performance (Wallace and Newbold, 1992). One of the major effects from the use of small quantities of yeast to improve performance in ruminant animals is on feed intake. For example, higher feed intake with the supplementation of yeast has been reported in calves (Fallon and Harte, 1987; Hughes, 1988), in steers (Adams, Galyean, Kiesling, Wallace and Finter, 1981) and in bulls (Edwards, Mutsvangwa, Topps and Paterson, 1990). Improvements in productive characteristics have further been reported by the use of yeast, for example, milk yield and composition in dairy cattle and goats (Hoyos *et al.*, 1987; Teh *et al.*, 1987; Williams and Newbold, 1990) and improved growth rates and feed utilisation in meat producing animals (Ruf *et al.*, 1953; Fallon and Harte, 1987; Hughes, 1988; Williams and Newbold, 1990) (see Table 8.12).

Table 8.12 *Some of the effects of yeast culture supplements on the activities of mixed populations of ruminal bacteria.*

Effect	*References*
Decreased ammonia concentration	Dawson and Newman (1987) Harrison *et al.* (1988)
Altered VFA production	Teh *et al.* (1987) Grey and Ryan (1988) Harrison *et al.* (1988) Williams (1989) Martin *et al.* (1989)
Increased ethanol concentration	Ingledew and Jones (1982) Bruning and Yokoyama (1988)
Moderated ruminal pH	Malcolm and Kiesling (1986) Teh *et al.* (1987) Williams (1989)
Decreased lactic acid concentration	Williams (1989)
Decreased soluble sugar concentration	Williams (1989)
Reduced methane production	Williams (1989)
Altered digestive patterns	Ruf *et al.* (1953) Gomez-Alarcon *et al.* (1987) Wiedmeier *et al.* (1987) Williams (1989) Chademana and Offer (1990)
Stabilised fermentation	Harrison *et al.* (1988)
Increased concentration of anaerobic bacteria	Wiedmeier *et al.* (1987) Harrison *et al.* (1988) Dawson *et al.* (1990)
Increased concentration of cellulolytic bacteria	Wiedmeier et al. (1987) Harrison *et al.* (1988) Dawson *et al.* (1990)
Increased concentration of yeast in the populations	Dawson et al. (1990)

(Dawson, 1990)

Aspergillus

While *Aspergillus spp.* have also been used, for example A. *niger* and A. *oryzae*, it is the latter that has received greatest attention. Quite substantial responses to supplementation with live A. *oryzae* plus the growth medium, have been reported, particularly with beef cows and calves on poor pastures. Growth rates of 800g/day have been achieved, compared with 570g/day for untreated animals (Wiedmeier, 1989). It has been suggested that research in dairy cows has tended to dwell too much on milk production with too little consideration of the associated changes in liveweight and body composition (Williams and Newbold, 1990).

Mode of action of fungi

The mode of action has not been clearly established but a model has been proposed by Dawson (1990) and the relationships between rumen metabolism and animal performance are given in *Figure* 8.6.

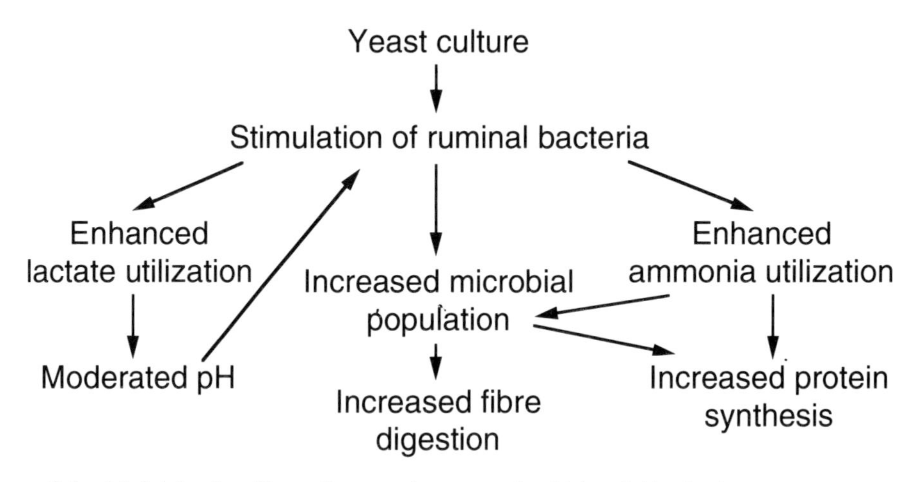

Figure 8.6 Model for the effects of yeast culture on microbial activities in the rumen (Dawson, 1993).

Fungal cultures have been reported to increase the total bacterial count in the rumen generally, and particularly, the cellulolytic population. For example, Harrison *et al.* (1988) found a 60% increase in total bacteria and an 82% increase in cellulolytic bacteria in the rumen of lactating dairy cows given 57g/day of yeast. *In vitro* increases of up to 461% in total bacteria have been shown, e.g. *Aspergillus oryzae* was reported to have increased the total bacteria by 12% and cellulolytic bacteria by 56% (Wiedmeier *et al.*, 1987) and by 80 and 188%, respectively (Frumholz *et al.*, 1989).

It appears that rate of fibre digestion is one of the major factors to be altered by the addition of, for example, yeast. It has been suggested that changes in the time course of digestion may increase the availability of nutrients in the rumen and have a significant effect on feed intake (Williams and Newbold, 1990; Dawson, 1993). Such increases in degradation rate would be particularly important for forages.

In ruminant animals the supply of amino acids to the small intestine is of major significance. As the rumen micro-flora influence this supply, there has been considerable interest in the so-called by-pass protein, i.e. protein that can reach the small intestine without being influenced by the rumen. A further approach is to manipulate rumen fermentation to enhance and modify the supply to the small intestine. It is generally accepted that the amino acid profile of the rumen bacteria is fairly constant. However, it has been shown that this can be modified by the addition of yeast culture and that the supply of the normally limiting amino acids to the duodenum (e.g. methionine and lysine) can be increased (Erasmus *et al.*, 1992).

Other important aspects of yeast could be:

Stabilisation of ruminal pH

This could explain the increased bacterial numbers, i.e. lowering pH is known to reduce growth of ruminal bacteria in a pure culture (Russel *et al.*, 1979). As an example, the yield of bacterial cells fell by 36% when the pH was reduced from 6.7 to 6.0 for mixed ruminal bacteria (Strobel and Russell, 1986). The elevated ruminal pH caused by *S. cerevisiae* is probably associated with the stimulation of *Selenomonas* which is a lactic acid mobilising bacterium. It is thought that the reduction in ruminal lactate is due, at least in part, to a reduced production caused by a lower substrate supply. In addition, there is an increased lactate uptake. These effects lead to a reduction in, and stabilisation of, rumen pH and consequently rumen fermentation.

Reduction in rumen ammonia

Substantial reductions in rumen ammonia levels (20 to 34%) together with increased microbial growth have been reported (Williams and Newbold, 1990). This might be explained by the uptake of ammonia by the yeast, so stimulating the microbial population.

Pattern of rumen fermentation

The pattern of rumen fermentation is changed in the presence of fungal cultures with volatile fatty acid production stimulated by both *A. oryzae* and *S. cerevisiae*. In the case of yeast, it tends to have no effect or causes a reduction in acetate to propionate ratio, probably through an increased propionate rather than a decreased acetate production. In contrast, *A. oryzae* has been reported to increase acetate to propionate ratio often with an increased butyrate production (Williams and Newbold, 1990).

Factors affecting the performance of fungal cultures

The effectiveness of fungi in dairy cow diets is related to the stage of lactation. Wallentine *et al.* (1986) and Gomez-Alarcon (1990) found the response to *A. orysae* inclusion to be greater in early as opposed to mid or late lactation. Similar results have been reported for *S. cerevisiae* (Gunther, 1990; Harris and Lobo, 1988). The diet has also been reported to influence the response to fungal cultures with the largest effect in milk yield when the ratio of concentrate to forage in the ration is increased (Huber *et al.*, 1985; Williams *et al.*, 1991). It has also been reported (Newbold, 1992) that in beef animals, better effects were seen with corn silage diets than with grass silage.

Yeast strains for specific functions

The effects of commercial yeast strains are most marked with mixed forage/concentrate diets, and increase fibre digestion and forage intake. On the basis of some of the characteristics already mentioned, yeasts are being screened in order to match them to specific dietary functions. For example, work at the University of Kentucky (Dawson, personal communication) has already identified two strains with commercial potential for ruminants. For high concentrate diets, yeast strain 8417 stimulates *Selenomonas* and increases lactate utilisation. Yeast strain 228 is used with high forage diets to ensure early colonisation with cellulytic bacteria and faster fibre digestion.

Yeast for non-ruminants

Non-ruminants also have considerable ability for fermentation (Cole *et al.*, 1968). It takes place mainly (about 80%) in the large intestine where there is less opportunity for absorption. The horse is one exception to this and it has been shown that digestibility of a diet of fescue hay and commercial pelleted feed was improved by supplementation with a yeast culture (*S. cerevisiae*). Significant improvements in the digestibility of dry matter, acid detergent fibre, neutral detergent fibre, phosphorus, magnesium, potassium, calcium and zinc were shown (Pagan, 1990).

Modified Yeasts

Chromium Yeast

Yeasts can be modified for specific purposes, for example chromium yeast (i.e. yeast with a high chromium content).

The role of Chromium

It is well established that Cr is an essential nutrient and beneficial effects of its supplementation have been shown in both children and farm animals. Central to its essentiality is its role as a Glucose Tolerance Factor (GTF). The classic work of Schwartz and Mertz in 1957 showed that the addition of brewer's yeast to the diet improved the rate of glucose removal and they named it the Glucose Tolerance Factor. In 1959, the same workers identified the GTF as involving trivalent Cr. Although its structure is not fully understood, it is now generally accepted that in addition to trivalent Cr, nicotinic acid, glutamic acid, cysteine and glycine are involved. The GTF apparently enhances the binding of insulin to cell receptors, thereby stimulating uptake of glucose by the tissues.

There is evidence that a marginal Cr intake is linked to elevated serum lipids in both humans and animals. For example, rats fed a low Cr diet had increased serum cholesterol, aortic lipids and plaque formation. The addition of 1 to 5mcg Cr/1 of drinking water reduced serum cholesterol. Athletes use chromium to build muscle mass and chromium yeast tablets are found in health food stores.

Sources of Chromium

Chromium occurs widely in nature and, as it exceeds 1/3000th of the earth's crust, there are only seven more abundant metals. It is a transition element that occurs in a number of oxidation states. The most stable form is the trivalent state which is the one involved in the GTF. Many forms of chromium are particularly inert and unavailable to the animal and in some cases toxic (hexavalent Cr). Natural sources of Cr (such as Cr yeast) are more bio-available than chromic salts, with improvements of up to eight fold. The initial identification of a factor in glucose tolerance was in brewer's yeast. Yeasts are able to incorporate Cr and provision of the correct fermentation medium results in a high Cr yeast which has been used in human nutrition and is now being used in animal nutrition.

Chromium in animal nutrition

Experiments in Louisiana State University (Page *et al.*, 1991) have shown improvements in carcass quality with the inclusion of 200ppb. Cr in the diet (in this case as chromium picolinate which is used in human nutrition and thought to be chemically similar to part of the GTF in Cr yeast). Not only was lean increased and fat decreased but serum cholesterol was also reduced. The benefits of the added Cr were substantial and the work has given impetus for further research to define more accurately the responses of pigs to dietary Cr. Legislation says little about Cr, presumably because there is not much information, but, at the levels used, evidence from human nutrition suggests no problems.

Selenium yeast

The requirements for selenium (Se) are intimately linked with those for vitamin E. Perhaps too much emphasis has been placed on Se toxicity with the result that too little has been focussed on requirements. Lack of knowledge on requirements is further complicated by the relationship between Se content of plant materials in the animal's diet

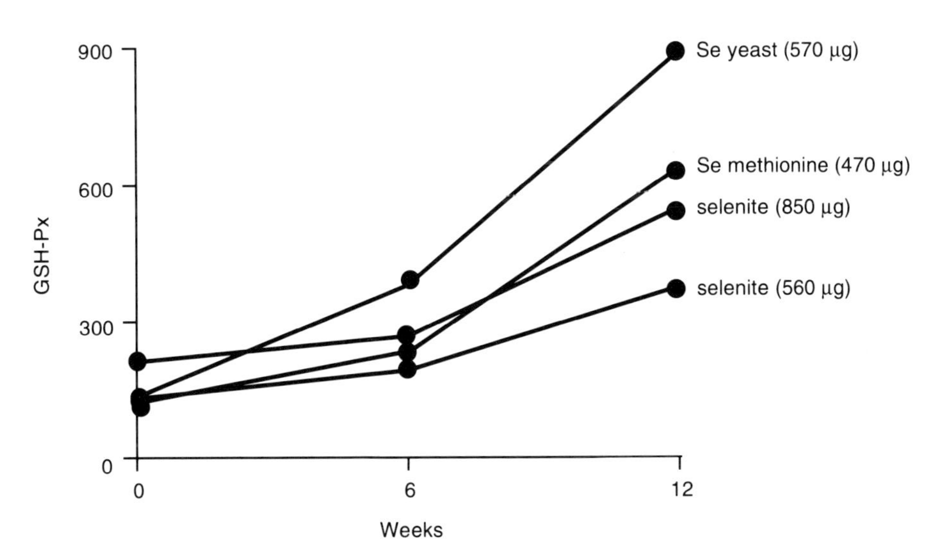

Figure 8.7 GSH-Px activity in erythrocytes (a measure of Se absorption) of heifers after dietary supplementation of a selenium deficient diet with different selenium compounds. The total daily amount of selenium fed is given in parentheses (Pehrson *et al.*, 1989).

and Se content of the soil. Se yeast may be used in place of the commonly used sodium selenite because of its claimed greater availability (*Figure* 8.7). Glutathione peroxidase (GSH-Px) contains 4g-atoms/mole and is often used as a measure of absorption of Se from different sources. Glutathione peroxidase in erytherocytes and other tissues of several species contains 4g-atoms/mole and probably plays an important role in destruction of peroxidases.

In many countries there is a desire to increase human dietary intake through levels in animal products (eg meat and milk). The animal itself requires adequate Se for many functions and recent interest has centred on stimulus of the immune system in the young.

Chemical probiosis

It is well known that the nature of the diet can influence the processes of digestion and absorption and that the conditions in the digestive tract and levels of *Lactobacilli* change after feeding. Chemical probiosis seeks to exploit this by providing dietary materials which positively influence gut micro-organisms and consequently the absorption of nutrients. In farm livestock, such techniques are used in improving productivity.

Organic acids

A good example of responses of this nature is in the young pig after weaning. This period of the pig's life is often characterised by poor growth, diarrhoea and mortality. The occurence of such problems is a function of time after weaning rather than chronological age or weight of the piglet. Typically such outbreaks occur about two weeks after weaning. The shedding of haemolytic *E. coli* is implicated in such problems and it has proved difficult to induce these conditions when the piglet is being suckled. A major difference in the suckled pig and the weaned pig is the lower levels of lactic acid in the gut of the latter.

Early attempts at chemical probiosis were set against this background of post-weaning conditions. They involved the use of organic acids which were often supplied in the drinking water in the belief that sick pigs would drink during a period of inappetance. Several acids proved beneficial in controlling haemolytic *E. coli* and improving growth performance but lactic acid was always predominant (Table 8.13). However, the effects of these materials were only evident during the period of administration.

Mode of action

Clearly the effects of the materials on the pH of the gastro-intestinal tract may have been a factor in the mechanism of response. However, it is interesting to note that in the work of Kershaw, Luscombe and Cole (1968) and Cole *et al.* (1968,1970) lactic acid was always more effective in the improvement of growth performance and it may well have its own specific effects.

Table 8.13 *Effect of lactic acid and propionic acid (0.8%) in the drinking water for 4 weeks after weaning on the* E. coli *population of the small intestine (million organisms/ml intestinal contents). For different sample litters the* E. coli *levels are given at weaning and then after treatment with lactic or propionic acid.*

	Litter	Pigs killed before treatment applied (i.e. at weaning)	Pigs killed after 4 week treatment peiod		
			Control	Lactic acid	Propionic acid
Dueodenum	1	444.20 (H)	112 (h)	0.30	1.00
	2	0.60 (H)	145 (h)	0.02	0.04
	3	0.60 (H)	2.70 (h)	0	2.00
	4	0.68 (H)	0.40 (h)	0	0
Jejunum	1	0.30 (H)	140 (h)	0.20	0.06
	2	14.50 (H)	2.30 (h)	0.04	0.10
	3	5.50 (H)	0.20 (h)	0	3.40
	4	6.00 (H)	1.50 (h)	0	0.40
Treatment period (Weeks 1–4)					
Feed intake (kg/day)			1.01	1.03	0.93
Growth rate (g/day)			377	409	355
Feed utilisation (kg feed/kg gain)			2.72	2.53	2.68

(Cole *et al.*, 1968)
H were haemolytic *E. coli* and h is a mixture of haemolytic and non-haemolytic strains of *E. coli*. For other pigs in which *E. coli* were found there were non-haemolytic strains only.

Acidification of diets

There has been recent interest in the acidification of diets for monogastric animals, for example the pig just after weaning. A mature pig is able to adjust stomach pH by secretion of hydrochloric acid from the parictal cells with very acid values being reached (as low as pH 2.0). Young pigs are different; while the newborn pig does produce some hydrochloric acid, its secretory capacity is severely limited. The consequence is that the young pig would have a pH of 4–7, much higher than the optimum for the enzymes. For example, pepsin has two pH optima, pH 2.0 and pH 3.5, and at higher levels protein digestion would be reduced. These effects are quite separate from the effects reported by Cole *et al.* (1968) of a large beneficial change in the bacterial flora as a result of the addition of 0.8% lactic acid to the drinking water. A number of organic acids have been used in the diet of weaned pigs, e.g. fumaric acid, citric acid, propionic acid. Generally good responses have been reported in terms of growth performance (Table 8.14).

Table 8.14 *Effect of graded levels of fumaric acid on growth performance of weaner pigs.*

	Fumaric acid level (%)				
	0	1	2	3	4
Diet pH	5.96	4.77	4.33	3.98	3.80
Liveweight gain (g/day)	261	261	257	296	297
Feed intake (g/day)	501	484	445	493	493
Feed:gain ratio (kg feed/kg gain)	1.92	1.85	1.75	1.67	1.67

(Easter, 1988)

Electrolyte balance

Any chemical that dissociates into its constituent ions is an electrolyte. A strong electrolyte will dissociate completely or nearly completely but a weak electrolyte will only dissociate to a limited extent (Table 8.15). Consequently, a solution of strong electrolytes will consist mostly of ions and in a weak solution there will be a large proportion of undissociated particles.

Table 8.15 *Some common electrolytes and non-electrolytes.*

Strong electrolytes	*Weak electrolytes*	*Non-electrolytes*
Hydrogen chloride	Hydrogen fluoride	Glucose
Sodium chloride	Ammonia	Sucrose
Sodium hydroxide	Acetic acid	Ethanol
Potassium fluoride	Mercuric chloride	Oxygen
	Acetone	

(Patience, 1989)

Individual electrolytes have specific roles. For example, sodium is found extensively throughout the body and has major roles in the maintenance of osmotic pressure and acid-base equilibrium, control of water metabolism and transportation of ions across cell boundaries.

A recent approach has been to consider electrolyte balance, because the balance of dietary cations and anions is closely related to performance in farm animals. There are two major estimates of electrolyte balance (Patience, 1989). These are dietary undetermined anion (dUA) and dietary electrolyte balance (dEB).

$$dUA = (Na^+ + K^+ + Ca^{2+} + Mg^{2+}) - (Cl^- + H_2PO_4^{2-} + SO_4^{2-})$$

$$dEB = Na^+ + K^+ - Cl^-$$

Both are expressed in millequivalents (mEq) or milliosmoles (mOsm) as it is the electrical and osmotic properties that are of interest. For example, in a feedstuff:

$$\text{dEB (mEq/kg)} = \frac{\text{Na(g/kg)} \times 1000}{23} + \frac{\text{K(g/kg)} \times 1000}{39} - \frac{\text{Cl(g/kg)} \times 1000}{35.5}$$

dEB is often used for convenience as it involves only three analyses and in many situations is useful. dUA is more accurate but is laborious.

Improved growth rate has been reported in weaner pigs when dEB was about 155mEq/kg (Cole *et al.*, 1992) (*Figure* 8.8). A value of 250 mEq/kg has been suggested for 4 week old chicks and there are well known effects of dEB on egg shell quality in laying hens (Mongin, 1981).

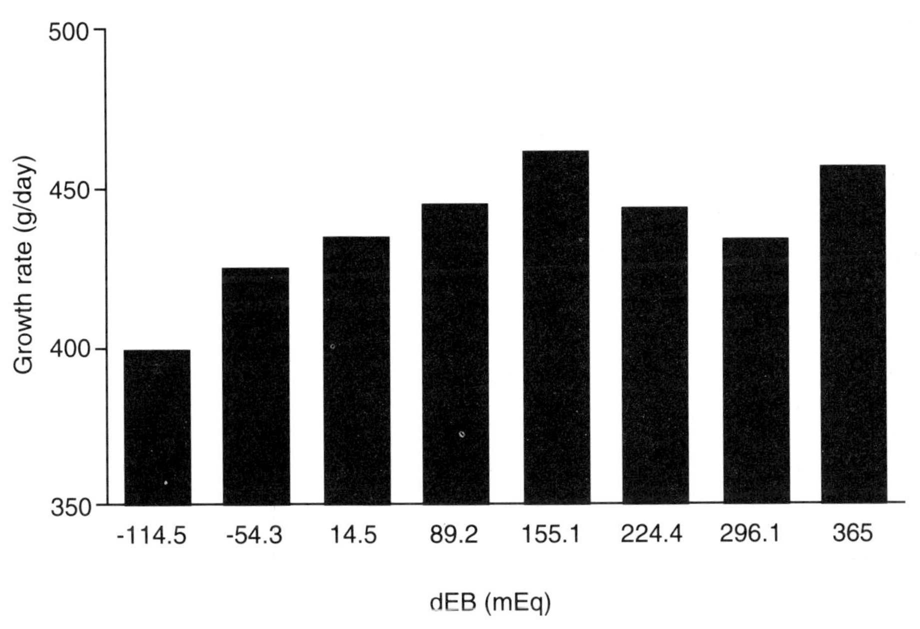

Figure 8.8 Electrolyte balance (dEB) and growth rate in pigs (8–30kg liveweight). (After Cole *et al.*, 1992)

Oligosaccharides – carbohydrate directed interactions

It is interesting that different bacteria attach preferentially in different organs. For example, while some settle in the digestive tract others may prefer the urinary tract or other organs. Consequently, cell recognition is important to bacteria. It is suggested that carbohydrates (sugars) are the primary markers for cell recognition with all cells carrying a sugar coat. The micro-organism has glycoproteins (lectins/fimbriae) on the surface of cells, which can recognise and combine rapidly/selectively and reversibly with the sugar (oligosaccharide) of the gut wall (*Figure* 8.9).

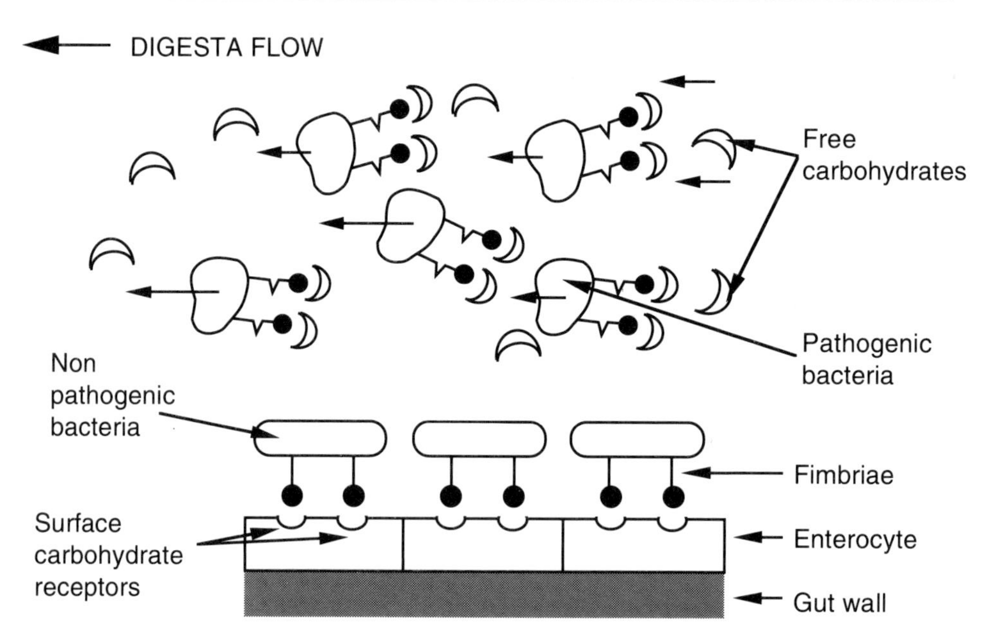

Figure 8.9 The lectin-carbohydrate combination is specific to a particular organism. However, if the same carbohydrate (e.g. an oligosaccharide) is provided in the diet, harmful bacteria can be encouraged to attach to these and they do not adhere to the gut-wall but are excreted without producing toxins.

Recently there has been interest in using this lectin/carbohydrate relationship in a probiotic way. Oligosaccharides particularly are being used. For example, it is thought that encouragement of bifidobacteria by the oligosaccharides of milk gives the suckling animal a measure of good health. After weaning there is a period of vulnerability. Oligosaccharides, for example mannan and fructose oligosaccharides, have been used in the diets of weaned pigs and poultry and are available in health food shops. Harmful bacteria must first of all attach to the gut wall to cause disease. If they attach to an oligosaccharide, not part of the gut wall, they pass out with the digesta (*Figure* 8.9) and do not cause a problem.

This adhesion has been defined for pathogens such as *E. coli*. They have common fimbriae (type 1) that recognise and bind to mannose receptors on enterocytes (Sharon, 1987). This adhesion can be reversed by the presence of D-mannose or D-mannopyranoside (–methyl-D-mannoside) (Ofek *et al.*, 1977). This has practical implications, for the addition of mannose (2.5%) to the drinking water of broilers has been shown to reduce significantly the level of *Salmonella typhimurium* and total bacterial levels in the bird's caecum (Oyofo *et al.*, 1989). –D-mannopyranose is found naturally in yeasts, e.g. *Saccharomyces cerevisiae*, as well as other mannans. It has been reported (Miles, 1993) that mannose can interfere with the attachment of *Salmonella*, *E. coli* and *Vibrio cholera* which have a mannose specific substance on the surface.

It has also been pointed out that carbohydrate directed interactions between cells are not restricted to pathological phenomena but are also crucially important in the healthy operation of the immune system (Sharon and Lis, 1993). These interactions have a role in directing leucocytes to specific parts of the body.

Combining oligosaccharides with a microbial inoculant has also proved effective in establishment of an intestinal ecosystem with a high 'barrier effect' against pathogens. (Mul and Perry, 1994).

Free oligosaccharides are natural constituents of food and feed. The term 'oligosaccharide' is used to cover substances consisting of between two and ten carbohydrate monomers. Their effect is structure (size, sugars and linkage) specific and dose dependent and influenced by the situation before their introduction. Their specificity relates to the many possible glyco protein-oligosaccharide interactions within the gut.

Fimbriae/lectin – protein directed interactions

The complete reverse of manipulating adhesion by the use of free oligosaccharides can also be applied. This is where bacterial attachment to the enterocytes on the gut wall is restricted by the addition of isolated fimbriae/lectins and their analogues (Pusztai *et al.*, 1980) (*Figure 8.10*).

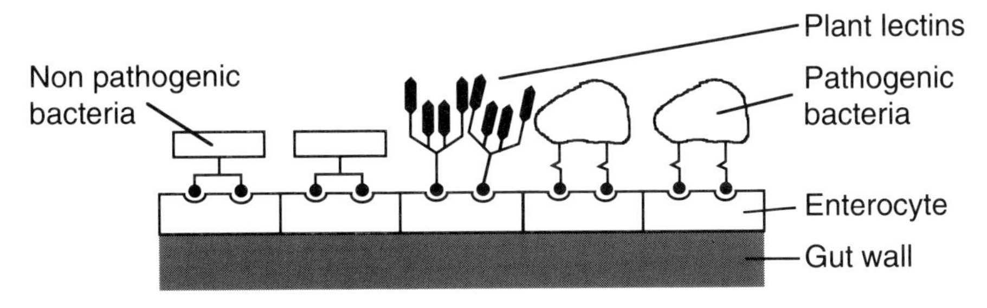

Figure 8.10 Competition by plant lectins for space on enterocyte receptors. The lectin occupies/utilises a binding receptor to the disadvantage of pathogenic bacteria.

Lectins are a class of protein that combines with sugars rapidly, selectively and reversibly. Like oligosaccharides, they are ubiquitous in nature and are found in plants.

Isolated fimbriae or plant lectins which have been present in the diet can occupy the binding site and prevent attachment of the bacteria. Dietary lectins may cause a change in the carbohydrate side chains of surface receptors of the small intestine brush border epithelium and, by creating favourable conditions for the attachment of selected bacterial species, lead to selective overgrowth. These principles show the flexibility open to nutritionists in beneficially controlling the micro-flora of the digestive tract.

Anti-microbial activity of plant extracts

Various plant extracts have been shown to be inhibitors of pathogenic bacteria, acting as a natural method to reduce their levels in the gastro-intestinal tract. One such extract is from freeze-dried garlic (*Allium sativum*) which has been shown to inhibit many bacteria, yeast, fungi and a virus (Rees *et al.*, 1993). This extract was found to be inhibitive towards *E. coli*, while having little effect on the beneficial *Lactobacilli* (Jezpura *et al.*, 1966; Rees *et al.*, 1993). Other inhibited bacteria include *Staphylococcus aureus* and *Candida albicans* (Huddlesaon *et al.*, 1944; Adetumbi *et al.*, 1988).

Non-viable cells

Work by Ewing and Cole (1988) with non-viable cells confirmed earlier work by Porter (1986) that it is possible to obtain growth benefits by feeding non-viable, non-pathogenic bacterial cells to animals. The growth enhancing effect of probiotics can, in some cases, be improved when the material is killed. Studies have been carried out in young pigs showing an improvement in liveweight gain of 15% (Table 8.16) and in young calves, improvements of 10% in liveweight gain and food conversion have been observed (Gribben and Hughes, 1989). Similar effects have also been seen in poultry (Sissins, 1988).

Table 8.16 *Effects of a processed bacterial culture in weaned pigs over a 4 week period.*

	Control	*Processed culture (2kg/tonne)*
Daily liveweight gain (kg)	0.26	0.30
% control	100	115.4
Daily feed intake (kg)	0.39	0.40
% control	100	102.6
Feed/gain ratio (kg feed/kg gain)	1.51	1.32
% control	100	87.4

(Ewing and Cole, 1988)

CHAPTER 9

Fermentation and industrial uses of micro-organisms

Introduction

This chapter deals with some of the basic principles of microbial fermentation and is intended to illustrate the background principles of commercial fermentation. An understanding of fermentation and growth of microbes in laboratory and commercial fermenters is useful in understanding the factors which will also affect their growth in the gut.

Fermentation - the stages involved

Micro-organisms are produced by fermentation which comprises a number of processes. These include:

- The formulation and development of the fermentation medium.
- The sterilization of the fermentation medium, fermenters and equipment.
- The production of a healthy typical inoculant.
- The production and multiplication of the organism in the fermenter leading to a bulking of biomass.
- The maintenance and control of fermentation conditions.
- The removal and collection of biomass from the growth medium.
- The drying and preservation of biomass into a stable form.
- Disposal of waste growth media and extraction of useful end products, e.g. enzymes.

Types of fermentation

There are two main types of fermentation, i.e. batch and continuous. The type used largely depends on the organism being produced and the capital and technology available.

Batch fermentation

The term batch fermentation is used for a closed fermentation system with no input or output of materials during the fermentation process. Growth takes place in stages, as the conditions are continually changing. Eventually a nutrient becomes limiting or a toxic metabolite develops, stopping or restricting growth. Most of the growth occurs when there is an excess of substrate and proceeds until the substrate is exhausted, or other changes in the medium become the limiting factor. The toxic products of metabolism, especially at higher substrate concentrations, can never be removed.

Fermentation goes through a number of stages (see *Figure 9.1*)

1. First, after inoculation, is the **lag phase**. No growth takes place at this stage as the organisms adapt to their new environment. Commercially the length of the 'lag phase' is kept to a minimum by using fast growing inocula, a balanced medium and optimum pH, oxygen levels, etc.
2. Following a period during which the growth rate of the cells gradually increases, they grow at a constant, maximum rate during the **log**, or **exponential phase**. In the exponential phase, the organism is growing at its maximum specific growth rate for the fermentation conditions.
3. Growth cannot continue indefinitely at an exponential rate. Increasing biomass results in the consumption of nutrients with substrate limitations and toxin production (or a combination of both). These all affect the organism and eventually the growth rate of the culture decreases until growth stops. The cells are then in the **stationary phase**.
4. The death phase takes place when lysis of cells occurs. At this stage the total quantity of biomass may decrease.

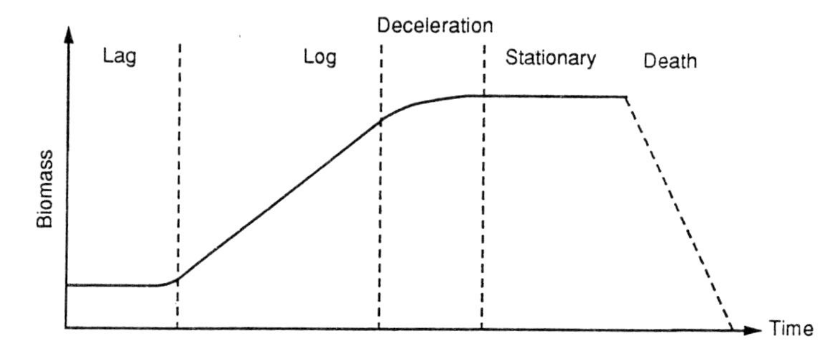

Figure 9.1 Phases in the growth of typical batch-grown microbial cultures. The slope and length of each section depends on the biomass being grown.

The lag phase

If fresh medium is inoculated with cells taken from a culture that has been grown into the stationary phase, a 'lag phase' occurs before growth starts. This is because cells in the stationary phase are changing their enzymic and chemical composition. If the inoculum is from a culture in the exponential phase of growth, the elapsed period for growth to restart will be much shorter and, at best, zero. However, if the medium is greatly different in composition then the lag phase may be prolonged, even if the cells are actively growing. This extended time for metabolic refinement can be seen with a medium in which two different sugars are present as energy sources. The organisms use one sugar first, then there is a short lag period after the first sugar is used and growth on the second sugar begins. This is called 'diauxic growth'. This period allows time for the synthesis or repression of enzymes or structural cell components and as a consequence it can be very short or very long.

Stationary phase

Unrestricted growth of a bacterial batch culture cannot continue indefinitely without its fermentation environment changing. The growing cell uses nutrients from the medium and also passes waste products into it. Eventually one or more of the nutrients becomes limiting or some waste product becomes toxic and growth is no longer possible. When the culture stops growing it enters the 'stationary phase'. The change from the exponential phase of growth to the stationary phase can be quite abrupt and is defined as the point when the culture depletes a limiting nutrient from a defined simple medium. If grown in a complex medium, the declining O_2 concentration, loss of nutrients, build up of toxic products and/or a shift to an unfavourable pH, slows and stops growth over a longer period. Bacterial cells have a complex mechanism for maintaining a constant intracellular environment even though concentrations of nutrients in the external environment are changing. When the limits of these conditions have been reached, growth rate slows. The total cell mass may stay constant but viability is likely to decrease and cell lysis may occur. Lysis of cells may lead to the production of complex products and secondary growth called 'cryptic' growth may follow.

Restarting growth

Adding sugar could re-start growth and give a higher biomass but, at some concentrations of glucose, another component of the medium will become limiting, e.g. nitrogen, magnesium, sulphur, oxygen supply, or pH. The restart of growth by immediate further addition to the substrate may occur, provided that the culture has not been in the stationary phase for too long.

Continuous culture

In batch type fermentation, exponential growth could, in theory, be maintained by the addition of fresh medium to the vessel. The medium must be formulated so that growth is substrate limited (i.e. limited by a component of the medium), and not toxin limited (ie. limited by toxic by-products). Exponential growth can, in theory, therefore proceed until the additional substrate has filled the vessels. If the medium added, displaces an equal volume of fluid from the vessel without losing sterility then this becomes a **continuous culture** for producing biomass.

Productivity in batch culture is best towards the end of the process, but in a continuous culture, running at the optimum dilution/addition rate under controlled conditions, productivity will be constant and always maximum. Thus, the productivity of the continuous system can be greater than the batch system. A continuous system can run for a very long time (weeks) so that the unproductive time (time for set-up, sterilisation, etc), is kept to a minimum. Batch cultures may be operated for only a limited time period and therefore the negative contribution of the set-up time is very high.

The productivity of biomass by a continuous culture, is therefore due to maintenance of maximum output throughout the fermentation and minimising of the non-productive period. In practice, the volumes involved with continuous fermentation are usually much smaller than batch culture owing to the increased labour requirement, etc. to keep them running. Continuous culture is an ideal method for the production of microbial biomass as the down time is low and productivity is high. Cell lysis is a problem which occurs in fast growing cultures. Nutrients should not be provided in excess, but added during the fermentation run in order to maximise biomass.

Contamination

As continuous fermentation can run longer than a batch fermentation, so the chance of contamination is much larger. If the conditions in a continuous culture are highly selective, then the contaminant, in order to become established, would have to be able to grow better in the environment than the culture organism. Thus, the contaminant would have to grow at a rate greater than the dilution rate taking place for the process. Contamination is generally a problem of poor fermenter design and construction, sterilization and operation techniques. If all these processes are kept under close scrutiny, contamination can then be kept to a minimum and controlling pH may also help.

Cell mutations

The probability of mutations occurring is greater in continuous than batch culture (Calcott, 1981) due to the increasing number of subsequent generations from the parent stock. As with contaminants, if the mutant growth is better adapted to the vessel conditions it may displace the original organism.

Isolation and screening of organisms to be used in fermentations

Isolation involves obtaining pure identified cultures, followed by their assessment to check for purity and type. The organism must be able to carry out the fermentation process economically and, therefore, selection of the culture for use is a compromise between productivity of the organism and economic constraints of the process. A number of criteria are thought important in the choice of organism (Bull *et al.*, 1979):

- The nutritional characteristics of the organism are important as different media will have different costs of producing biomass.

- The optimum temperature for growth of the organism must be considered because an organism with an optimum temperature for growth above 40° reduces costs associated with keeping the vessel cool, as heat is produced during fermentation.
- The organism should not react with the equipment used.
- The organism should be stable and easy to manipulate genetically.
- The organism should be as productive as possible.
- The organisms should be easy to remove from the medium.

Fermenter design

A fermenter is designed as a controlled vessel to provide optimum conditions for the growth of a micro-organism for biomass or product.

Vessel type and size

Preliminary laboratory trials are normally conducted in a shake flask (0.5 to 2 litres) while the culture volume in large commercial fermenters can be in the region 25,000–1,000,000 litres (see Table 9.1).

Table 9.1 *Fermenter size.*

Vessel	*Volume (litres)*
Shake flasks	0.5 – 2
Laboratory stirred fermenters	5 – 20
Pilot-scale fermenters	50 – 5,000
Production fermenters	25 – 1,000,000

Shake flasks

This method of culturing micro-organisms allows the investigation of a large number of experimental variables in a short time with low expenditure (Plate 9.1). Optimum conditions for product formation are often established on this scale.

Laboratory stirred fermenters

Scaling-up a shake flask fermentation into a stirred fermenter is normally the second stage. It allows the testing of certain parameters which cannot be conveniently examined on the shake flask scale, for example, the effects of pH, aeration-agitation, etc. (*Figure* 9.2).

Plate 9.1 Laboratory shake flasks. [*By courtesy of Kirk Robinson*]

Plate 9.2 Typical laboratory fermenters. [*By courtesy of Kirk Robinson*]

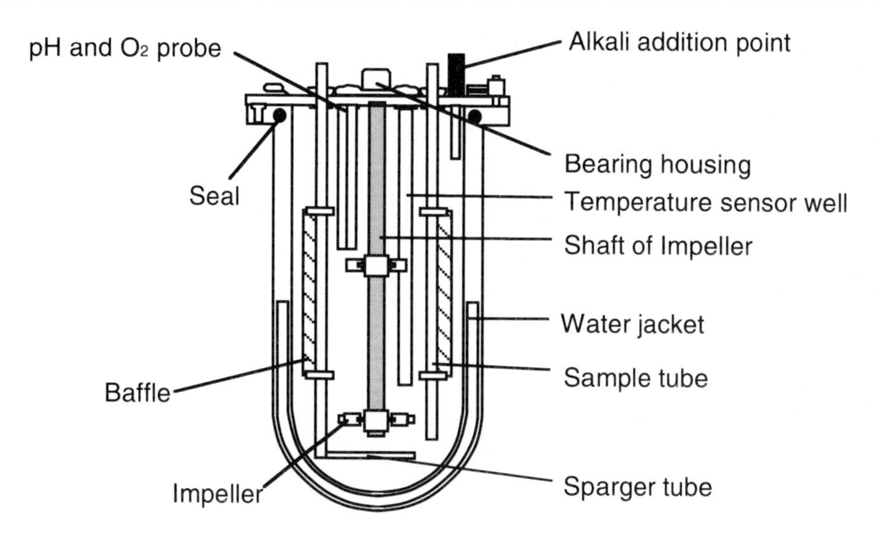

Figure 9.2 Components of laboratory fermenter.

Functions

Bearing housing	–	Must be air-tight and not allow entry of contaminated air.
Seal	–	Secures joint with sides and top.
Temperature sensor well	–	Allows a probe to monitor medium temperature.
Shaft of impeller	–	Drive impellers.
		Water jacket maintains constant internal temperature.
Sample tube	–	Allows samples to be taken to monitor growth.
pH and O_2 probe	–	Allows the state of the fermentation to be recorded and can be used to affect alkali addition and oxygen transfer into the vessel.
Sparger tube	–	Sterile air is injected through this tube which is generally made up of small holes. This ensures better aeration.
Baffle	–	Improves mixing and aeration.
Impeller	–	Mixes and aerates.

Pilot-scale fermenters

The medium for shake flasks and laboratory fermenters is normally sterilised by autoclaving, whereas in commercial pilot and production scale fermenters, the medium is sterilised by live steam injection.

Production fermenters

Production fermenters can be in the size of 25 to 1,000,000 litres depending on the organism being grown (*Plate* 9.3 shows the top of a typical commercial fermenter for growing yeast cultures). The medium is normally sterilised 'in line' rather than in the fermenter as it would not be feasible to bring the entire vessel and contents to a sterilisation temperature and pressure.

As fermenters are scaled up, it is important to note that oxygen transfer rate is inversely related to the liquid volume/flask volume ratio.

Plate 9.3 Top view of a commercial fermenter. [*By courtesy of Kirk Robinson*]

Requirements of a fermenter

The important parameters affecting fermenter design are:

- Capability for aseptic operation.
- Good oxygen transfer and supply to microbial cells with high demand.
- Precise temperature control, with removal or addition of heat.
- Bulk and uniform mixing.
- Few joins and corners in vessel and pipework.
- Easily usable sampling ports.

Heat production during fermentation

Heat is produced when microbes ferment a substrate and also as a result of agitation, etc. If the heat produced is limiting the growth of that organism or makes growth less cost effective it will have to be removed, normally by the use of a cooling coil or jacket. In small scale laboratory fermenters the vessel can be cooled by placing it in a water bath. It is usual to put cooling coils in a fermenter greater than 500 litres (Muller and Kieslich, 1966).

Aeration and contamination

Sterile air is required in very large amounts in many aerobic fermentations, e.g. *B. subtilis*, and care must be taken to have an adequate sterile filtration system. Filters must be carefully sterilized prior to use and be effective in sterilizing the incoming air. Air can be passed through a series of filters from cheap glass wool to fine bacterial filters. The use of pre-filters increases the life of the final filter and reduces the possibility of blockage during a fermentation. It is important that a consistent volume of air enters the vessel to avoid oxygen starvation and therefore a large surface area of filter is needed.

Contamination during fermentation

Contamination during fermentation may be avoided by simple precautions which include:

Sterilising the vessel

The vessel can be sterilised by the injection of live steam, conventionally held at pressure for sufficient time to kill all the micro-organisms (*Figure* 9.3). In a laboratory fermenter this is 121°C and 15 psi for 15 minutes. Larger vessels are usually held at lower temperatures and pressures for shorter periods of time. As the vessel cools it is important that air entering it is filtered to avoid contamination. NB. If air is not allowed back into the vessel, a vacuum will occur, possibly causing it to collapse.

Inoculum and contamination

The inoculum is a potential source of contamination as it only takes one bacterium to gain entry by careless procedures, such as touching a stopper in the laboratory, poor inoculation techniques, or allowing contaminated air to come into contact with the growth medium, to destroy a fermentation run.

Ports

All ports are potential contamination points. The seals around them should be kept tight during fermentation. Sterile samples should be taken by using positive pressure and if necessary through flamed ports.

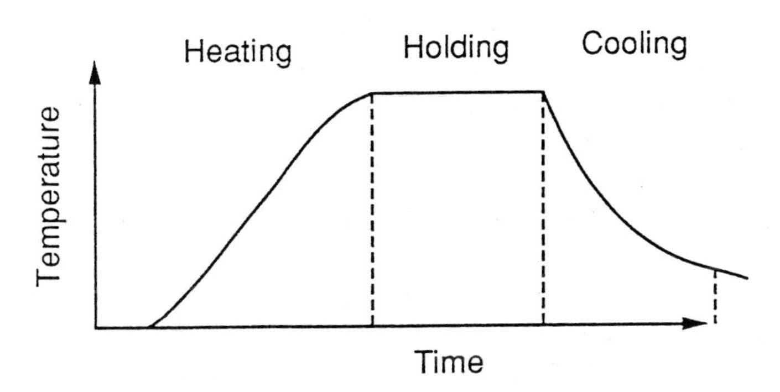

Figure 9.3 Sterilisation time-temperature profile. The fermenter contents are taken to the required sterilisation temperature and held at this temperature for a set period. The vessel and contents are then cooled to the optimum temperature for the growth.

Antifoam and acid/alkali as a source of contamination
Especially with antifoams, the liquid entering should be pre-sterilized before addition to the main vessel. Where extremely strong acids or alkalis are used for pH control, there is no need to autoclave prior to addition.

Storage of cultures

Agar slopes with suitable organisms may be stored in a fridge or freezer and need regenerating at approximately 6 monthly intervals. By using very low temperatures, the metabolic activities of micro-organisms may be reduced considerably, e.g. with the use of a liquid nitrogen refrigerator. Good storage by this technique involves growing a culture to the stationary phase, re-suspending the cells in 10% glycerol and freezing the suspension in sealed ampoules before storage under liquid nitrogen. The best results can be achieved by freezing the suspension quickly and thawing rapidly to retrieve the culture. Cultures are frozen and are kept at ultra-low temperatures (–82.5°C). It is important that the frozen concentrates are not allowed to melt and refreeze. Ice crystals form during freezing and can damage the cell membranes of bacteria. All preserved cultures should be checked routinely for quality and viability by a procedure such as that outlined by Lincoln (1960).

Medium formulation

The design of a good medium for bacterial fermentation is as important a stage in producing biomass as feed formulation is an animal nutrition (Tables 9.2 and 9.3). In general terms the chemical composition of all bacteria is very similar, yet there is a wide variation in the basic nutrients required by different species. The difference in requirement for cultivation *in vitro* reflects the natural environmental adaptations of the different species and thus their varying abilities to synthesise materials. The different constituents of the medium must meet all the energy and nutrient requirements for cell maintenance and synthesis (*Figure* 9.4 illustrates the response of increased substrate on cell growth). The mineral requirements of the media can, however, often be determined by looking at the composition of the cell. Typical nutrient levels in cells and media are shown in Tables 9.4 and 9.5 respectively.

All microbes require a supply of water to grow. If the water potential of the surroundings is less than that of the microbes, they will lose water and die. A higher water potential will cause them to absorb too much water and burst. Careful control of water potential is therefore essential for microbial growth. Most commercial fermentations, however, take place in a liquid broth with extreme water potentials unlikely to occur.

In addition to carbon, hydrogen, oxygen and nitrogen (which are the main elements necessary for the growth of bacteria) other materials are required in smaller amounts, e.g. phosphorus, calcium, magnesium, potassium, sulphur, sodium, iron, etc., (see Table 9.6). Also, some essential metabolites are required in almost infinitesimal quantities since

Table 9.2 *Typical medium for the growth of* Lactobacilli.

1.	MRS Broth (pH 6.2)	
	Peptone (g)	10
	Lab-Lemco powder (g)	8
	Yeast extract (g)	4
	Dextrose (g)	20
	Tween 80 (g)	1
	Di-potassium hydrogen phosphate (g)	2
	Sodium acetate $3H_20$ (g)	5
	Tri-ammonium citrate (g)	2
	Magnesium sulphate $7H_20$ (g)	0.2
	Manganese sulphate $4H_20$ (g)	0.2
	Distilled water (g)	1000
	Sterilize by autoclaving at 15 p.s.i. for 15 m at 121°C	

2.	MRS Agar
	MRS Broth + 2% 'ionagar' which is sterilized by autoclaving at 15 psi for 15 m at 121°C

(de Man, Rogosa, Sharpe, 1960)

Table 9.3 *Medium for growth of* Lactobacillus bulgaricus.

Component	*Quantity*
Casein peptone, tryptic digest (g)	10.0
Meat extract (g)	10.0
Yeast extract (g)	5.0
Glucose (g)	20.0
Tween 80 (g)	1.0
Di-potassium hydrogen phosphate (g)	2.0
Sodium acetate	5.0
Di-ammonium citrate (g)	2.0
Magnesium sulphate (g)	0.2
Manganese sulphate (g)	0.05
De-ionised water (mg)	1000.0
pH 6.2 – 6.5	
Potential yield	5g/l

(McEntree, 1989)

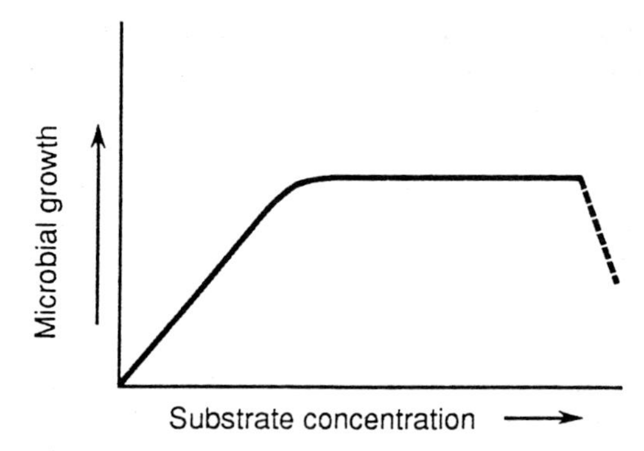

Figure 9.4 Effect of substrate concentration on the rate of microbial growth. Growth is seen to increase quickly initially, but further addition has no effect. Eventually high substrate concentration can limit growth.

Table 9.4 *Typical composition of micro-organisms (% of dry weight).*

Organic				
	Bacteria		*Yeasts*	
Component	*Average*	*Range*	*Average*	*Range*
Carbon	48	46–52	48	46–52
Nitrogen	12.5	10–14	7.5	6–8.5
Protein	55	50–60	40	35–45
Carbohydrate	9	6–15	38	30–45
Lipid	7	5–10	8	5–10
Nucleic Acid	23	15–25	8	5–10

Inorganic		
	Bacteria	*Yeasts*
Phosphorus	2.0–3.0	0.8–2.6
Sulphur	0.2–10.0.	01–0.24
Potassium	1.0–4.5	1.0–4.0
Magnesium	0.1–0.5	0.1–0.5
Sodium	0.5–1.0	0.01–0.1
Calcium	0.01–1.1	0.1–0.3
Iron	0.02–0.2	0.01–0.5
Copper	0.01–0.02	0.002–0.01
Manganese	0.001–0.01	0.0005–0.007
Molybdenum	–	0.0001–0.0002
Total ash	7–12	5–10

(Buckland, 1988)

Table 9.5 *A typical range of concentrations of mineral components used in fermentation (g/l).*

Component	*Range*
KH_2PO_4	1.0–4.0
$MgSO_4 .7H_20$	0.25–3.0
KCl	0.5–12.0
$CaCO_3$	5.0–17.0
$FeSO_4 .4H_20$	0.01–0.1
$ZnSO_4 .8H_20$	0.1–1.0
$MnSO_4 .H2$	00.01–0.1
$CuSO_4 .5H_20$	0.003–0.01
$Na_2MnO_4. 2H_2O$	0.01–0.1

(Buckland, 1988)

Table 9.6 *Explanation of typical fermentation media ingredients.*

Component	*Method of supply*
Carbon:	Glucose Starch or flour Vegetable oils Molasses Lactose Sucrose
Nitrogen:	Ammonium (wherever possible; sometimes ammonia gas is used for controlling pH). High protein sources, e.g. soyaflour, corn gluten, cottonseed flour (these typically contain 50% protein). Specialised nitrogen sources, e.g. cornsteep liquor, yeast extract. (These are more expensive and are usually used in smaller quantities to provide a specific component or to provide the right balance between available nitrogen and insoluble nitrogen, or to provide key trace elements).
Phosphorus:	Inorganic phosphate – often used in low concentrations due to inhibition of antibiotic synthesis. Organic phosphates, such as those found in cornsteep liquor, often work well because they provide a slow release of phosphate.
Calcium:	Calcium carbonate is often an important ingredient. Supplies calcium to the cell as well as acting as a buffer. Calcium forms an insoluble precipitate with certain antibiotics.
Magnesium:	Usually supplied as magnesium sulphate and materials such as cornsteep liquor.
Potassium:	Cell requirements for this are large.
Trace Metals:	Usually supplied through other raw materials.

they function essentially as catalysts. These essential metabolites, in many instances, have a similar role to the vitamins required in human nutrition.

The medium should:

- Produce the maximum yield per gram of substrate in the quickest time.
- Produce minimum by-product.
- Be cheap and consistent.
- Should not affect pH, foaming, shape or composition of the organism.

Crude medium constituents may give rise to scale-up problems due to their activity in chemical reactions during sterilisation of the medium. Fermentation medium is subjected to more severe heat on large scale batch sterilisation, due to the prolonged period needed to bring the larger medium volume to the required sterilisation temperature and also the increased length of time taken for the large mass to cool. Heating and cooling periods of 3–6 hours are not uncommon in large vessels. As stated earlier this can be overcome by sterilising both the vessel and the medium (in line) with live steam.

Many of the natural carbon and nitrogen sources contain all of the vitamins needed, at levels in excess of bacterial requirements. Yeast extract is a particularly good source of vitamins.

Specific requirements of *Lactobacilli*

Media have been described for the cultivation of *Lactobacilli* (Briggs, 1953; Cox and Briggs, 1954; Rogosa *et al.*, 1951a,b), to which various other ingredients have been added (Tables 9.2 and 9.3). For example, Tween 80 is a polyoxethylene sorbiton mono-oleate which improves growth rate (Briggs, 1953) as does acetate. Manganese is needed in excess of its demand for cell structure, compared with other micronutrients (Orla Jensen, 1943; Macleod and Snell, 1947), and minor components (e.g. citrate) often having a significant effect on yield.

Buffering of media

This is often done by the inclusion of calcium carbonate (as chalk) giving a neutral pH. If the pH decreases, the carbonate is decomposed and hydroxide is released restoring the pH to neutral, if needed. Phosphates are necessary in the media as phosphorous is part of the cell walls and bacterial nucleic acids. They are also involved in buffering.

Fermentation conditions

Oxygen

This is one of the most important factors affecting cell growth, particularly in aerobic fermentation. The oxygen need of a culture is affected by the nature of that medium, with more reduced carbon sources having a higher oxygen demand. Nutritional factors and bacterial type can influence the oxygen demand of the culture.

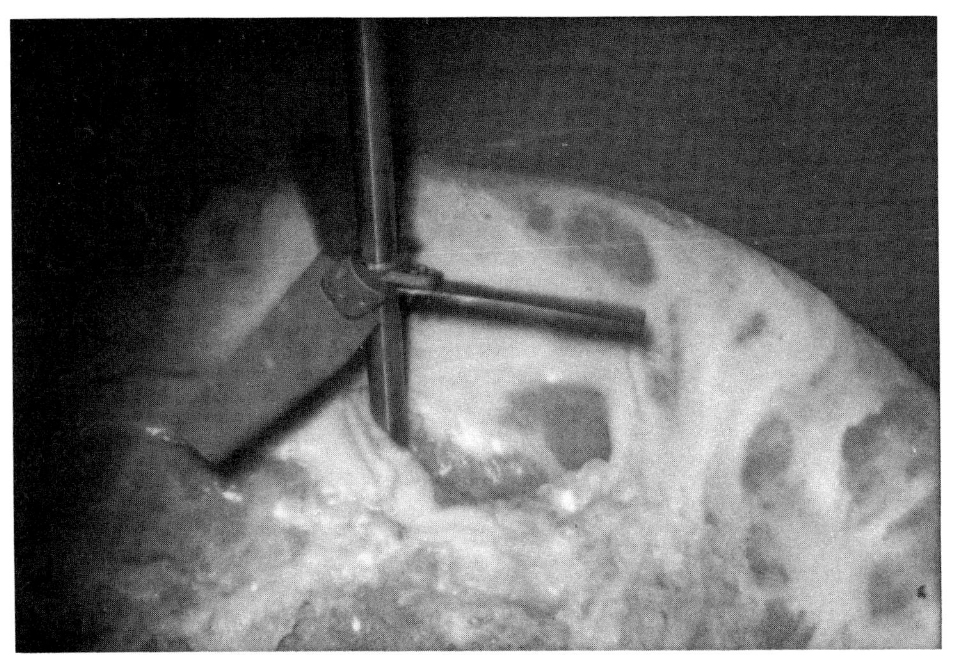

Plate 9.4 A typical agitator in a commercial fermenter. [*By courtesy of Kirk Robinson*]

Fermenter design and dissolved oxygen concentration

Impeller speed is one of the crucial factors in determining dissolved oxygen concentrations. In fermentation procedures it is generally found that impellers in smaller vessels are rotated at higher speeds than those in large vessels. The impeller tip speed is, however, maintained at near constant values (150–400m/min). Laboratory fermenters will normally be agitated at 100 – 500 rpm. Agitation of a fermentation is to dilute bacterial toxins from specific regions, maintain homogenous conditions in all parts of the fermenter as well as improving the oxygen transfer and dissolved oxygen concentration (see *Plate* 9.4).

Baffles are used to increase the dissolved oxygen concentration on agitation (*Figure* 9.5), and there is a wide range of agitators available for every fermentation situation (*Figure* 9.6).

Commercial air-lift fermentions now rely on agitation of the growth medium by streams of air passing through it. Two simple designs of air lift fermenters are shown in *Figure* 9.7. This is an effective way of dissolving oxygen and maintaining agitation.

pH control

pH is controlled in most fermentations either by means of a buffer as described earlier or a pH control system by the additon of acid or alkali solutions (e.g. ammonia or sodium hydroxide and sulphuric acid).

pH control is essential in large scale fermentations and is normally adjusted by automatic addition of ammonia, or sodium hydroxide and sulphuric acid. pH drift, due to acid hydrolysis is not uncommon, particularly when sugar-containing media are

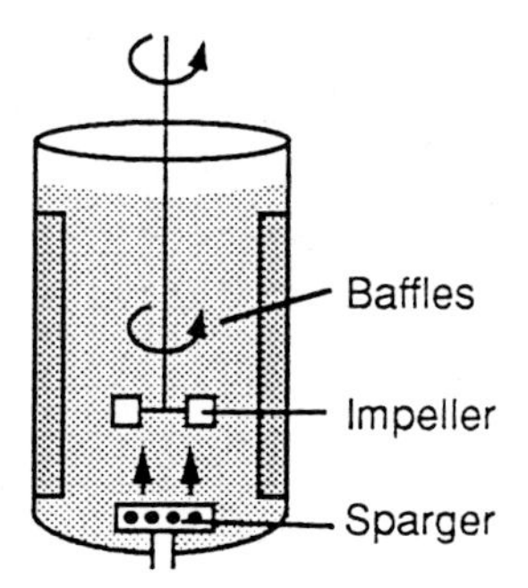

Figure 9.5 Typical design of commercial fermenter with central agitation and baffles.

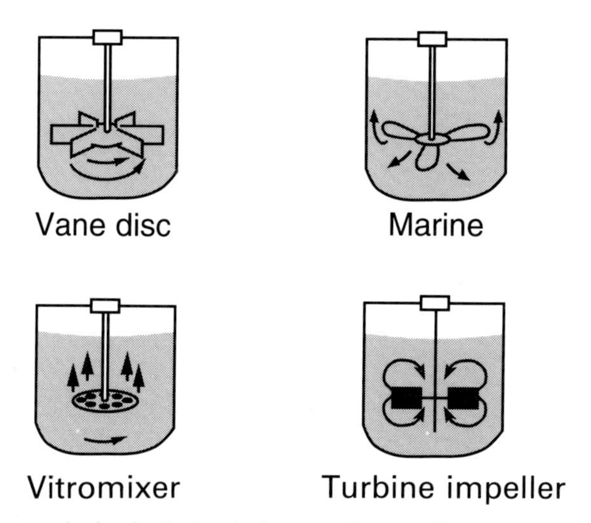

Figure 9.6 Different methods of agitation in fermentation vessels.

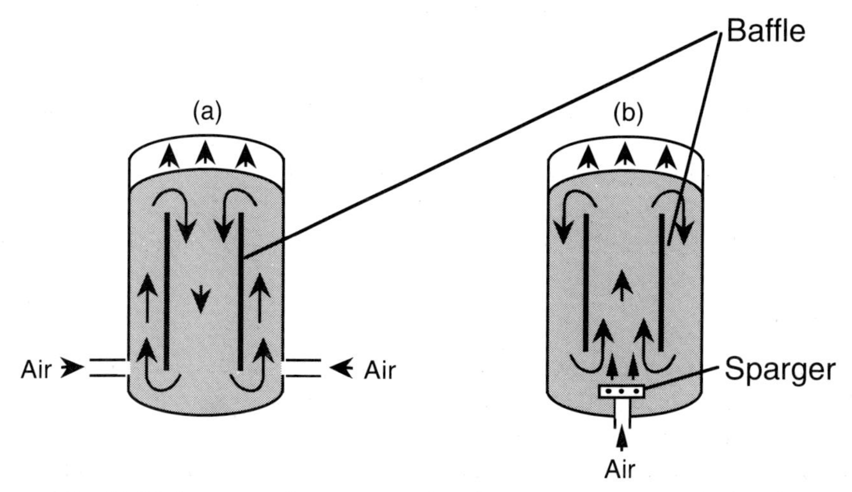

Figure 9.7 Air lift fermenters showing air and medium movement. Type (a) has the air feed from the sides and type (b) has the air feed from the base.

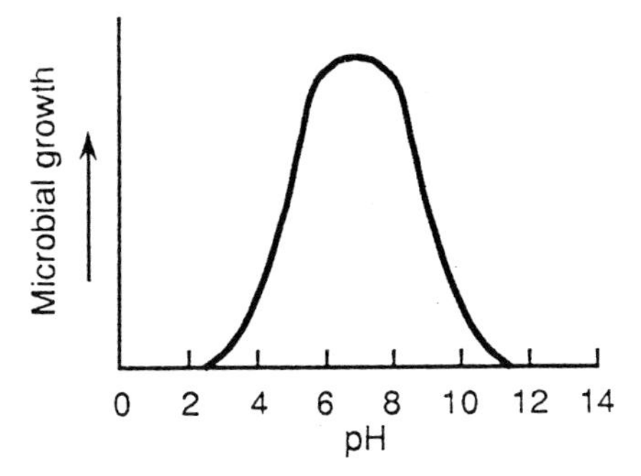

Figure 9.8 The effect of pH on microbial growth.

overheated or superheated in concentrated solution. Acid pH drift in these circumstances can lead to precipitation of mineral components.

There is an optimum pH for the growth for all microbes, with levels below or above this possibly reducing growth (*Figure* 9.8). Bacteria usually grow in the range of pH 5–8, yeasts 3–6, moulds 3–7, and higher eukaryotic cells 6.5–7.5. The pH will, however, change for different reasons during fermentation, e.g. if ammonia is the nitrogen source, the pH will fall. Ammonia in solution leaves a H^+ in the medium when it is used. If nitrate is the nitrogen source, then hydrogen ions are removed from the medium to reduce the nitrate to $R\text{-}NH_3$ and the pH tends to increase. Organic amino compounds also cause the pH to drift upwards as the compounds are deaminated. pH can be affected by the end products of fermentation, e.g. lactic acid, acetic acid, or pyruvic acid. In addition different bacteria have different optimums which affect their ability to colonise different surfaces within the gastro-intestinal tract.

Precipitation

One of the most common causes of precipitation in media, following heating, is the reaction between di- and tri-valent metals and soluble phosphates in the medium. The presence of either of these reactive groups may be intentional, i.e. as deliberate ingredients. Divalent metals such as calcium and magnesium, also precipitate in the presence of carbonate, the latter arising from the breakdown of bicarbonate buffer in the medium.

Foaming during a fermentation

Foaming is a common problem in industrial fermentation and may result in the loss of cells and blockage of exhaust filters. Foaming should, therefore, be kept under control either mechanically, chemically or else by the use of a non-foaming strain. Foaming which occurs early in fermentation is usually due to a component in the medium whereas, foaming later, is often related to a property of the fermentation organism. Foaming is controlled by the addition of an anti-foaming agent when levels become too high. Anti-foaming compounds are surface active agents which reduce the surface tension of the air bubbles in foams (Whitaker and Stanbury, 1984), they act competitively by replacing the compounds which cause the foaming (Solomons, 1967). Anti-foaming compounds include: alcohols, silicones, esters, fatty acids and their derivatives (e.g. soya oil, olive oil), and sulphonates (Solomon, 1969).

Inorganic ions for bacterial growth

All bacteria require the medium to supply inorganic ions for different essential functions. A list of the common ions is given in Table 9.7.

Table 9.7 *The use of inorganic ions in the bacterial growth medium.*

Inorganic ion	*Use for bacterial growth*
Ca^{2+}	Enzyme cofactor
Cl^{-}	Not often required
Co^{2+}	Found in vitamin B_{12} and derivatives
Cu^{2+}	Found in some enzymes
Fe^{2+}	Found in some enzymes and also in cytochromes
K^{+}	Important cation for internal cell balance and also found in cofactor of some enzymes
Mg^{2+}	Cofactor for enzymes
Mn^{2+}	Cofactor for enzymes
Mo^{2+}	Found in some enzymes
NH_4^{+}	Main structure in which inorganic N is used
PO_4	Takes part in many metabolic reactions
SO_4	Supplies S to growing cell

Inoculum

The inoculum used to start the fermentation should be pure, healthy, viable and active. As discussed earlier, this will minimize the length of the lag phase after its addition. The inoculum must be able to supply sufficiently large volumes to be of optimum size and should be a typical organism (Meyrath and Suchanek, 1972).

It is common to use large volumes of inoculum on a production scale (for example, 10% v/v), in order to reduce the initial growth phase of fermentation and thus make better use of available plant capacity. In order to provide these large volumes of inoculum, it is necessary to conduct seed-stage fermentations in the production plant (see Plate 9.5) and, thus, introduce additional stages into the inoculum development procedure. The introduction of these changes into the procedure, frequently causes scale-up problems due to culture degeneration or the altered metabolic state of the organism at the time of inoculation. The more stages between the stock culture and the production vessel, the greater the risk of contamination or even strain degeneration. A long lag phase is disadvantageous as medium is consumed in maintaining a viable culture without producing biomass and also wasting fermenter time. In order to avoid this, bacterial inocula should be transferred in the log phase of growth, when the cells are still metabolically active (Lincoln, 1960).

Plate 9.5 A typical seed vessel for a commercial fermenter. [*By courtesy of Kirk Robinson*]

Temperature and growth

Temperature can greatly affect growth (*Figure* 9.9) with all organisms having an optimum, maximum and minimum. With increasing temperatures growth rate increases until the optimum for that particular organism is reached. A few degrees around the optimum may not dramatically reduce growth or yields. Further increases above the optimum can damage enzymes within the organism and eventually lead to its death.

Examples of temperature range and optima are:

- Psychrophilic species: 0–25°C optimum 20°C
- Mesophilic species: 20–45°C optimum 25–37°C
- Thermophilic species: 45–60°C optimum 50–55°C

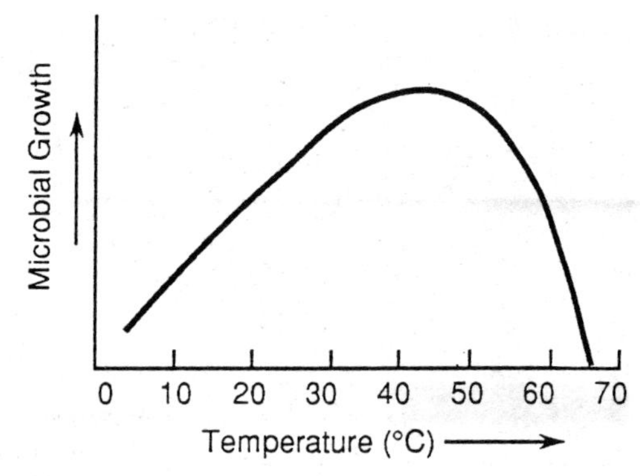

Figure 9.9 The effect of temperature on microbial growth.

Most bacteria and fungi belong to the mesophilic group.

Almost all bacteria which are pathogenic to man have an optimum temperature for growth of 37°C and the same is true of other animals where pathogenic bacteria have adapted to grow at body temperature. The physiological and biochemical aspects of the temperature responses of micro-organisms have been in dealt with in some detail by Ingraham and Stokes (1959) and Ingraham (1962). Temperature is known to affect cell size and morphology of some micro-organisms, but the effects are not well documented.

The fermentation process

Fermentation is a complex process of metabolism which depends to a great extent on bacterial nutrition (*Figure 9.10*) and the other conditions that have been outlined. Knowledge of these processes is exploited in commercial fermentation. The entire fermentation process from stored culture samples to isolation of biomass from culture broth is illustrated in *Figure 9.11*. The final step in this process is to remove the biomass from the medium usually by centrifugation (see *Plate 9.6*).

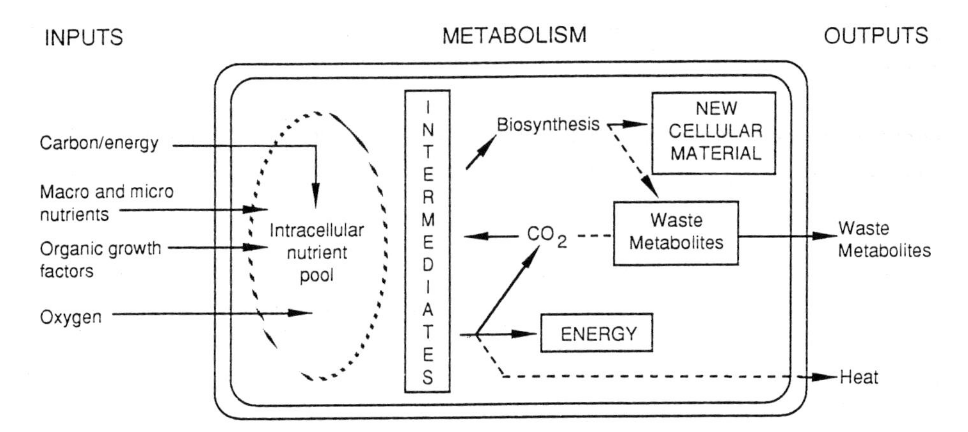

Figure 9.10 Illustration of components of basic bacterial nutrition and metabolism.

Plate 9.6 A commercial scale centrifuge [*By courtesy of Kirk Robinson*].

Measurement of cell number and biomass yield

Determination of biomass yield is essential to monitor the results of any fermentation. The following are examples of the many methods practised.

Direct microscopic counting. This is the determination of total cell number directly, by counting under a light microscope. This is carried out with a graduated slide called a haemocytometer.

Viable plate count. This method involves making known dilutions of the samples and plating them on Petri dishes with agar containing appropriate growth substrates. The agar is spread with diluted samples of micro-organisms either by hand or by spiral plating. Every colony produced originates from one organism providing the dilution is correct. (*Plate* 9.7 shows typical streaked agar plates).

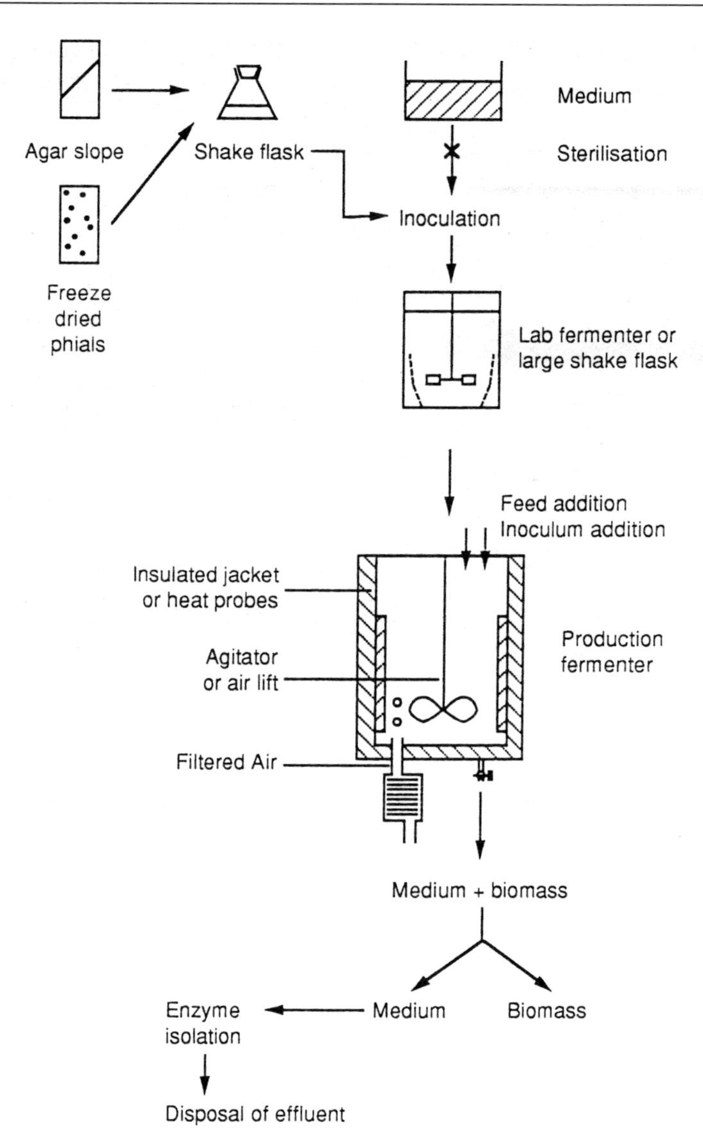

Figure 9.11 Flow diagram of simple fermentation process.

Coulter counter. A diluted culture sample is mixed with electrolyte and placed in the inner tube. A small volume of liquid is removed through a small hole of known size to create a vacuum. An electrical potential is then applied across electrodes with the size of the resistance change being related to the cell's size. The smallest size of hole is usually 30μ. The process is only suitable for single cells in liquid media.

Turbidity. The optical density of a cell sample is directly proportional to the size of biomass. It is a simple method, providing a fast and a cheap way of predicting biomass levels throughout a fermentation.

Plate 9.7 Hand streaked agar plates [*By courtesy of Kirk Robinson*].

Cell dry weight. This is the simple method of determining the quantity of cell biomass in a given sample. Samples are taken from the fermenter, centrifuged, washed with a buffer solution, and then dried for approximately 8 hours at 110°C. This technique is only useful for media which do not contain solids, as it is impossible to determine the quantity which is not biomass. If calcium carbonate has been used as a buffer, an acid wash should be used to dissolve it and remove it from the sample before drying. Cell weights can be under-estimated when water soluble components leak out of cells on washing. Drying may also decompose cell components.

Cell component as an indicator of yield. This method is useful where there are other semi solids in the substrate and a direct analysis cannot be made. Cell levels can be measured by a cell component (e.g. protein, RNA, DNA). The protein content of the cells is usually quite constant, while RNA varies largely with growth rate. DNA is probably the most consistent cell component which is not normally associated with non-cellular materials added as nutrients. DNA measurement however is time consuming.

Product formation. Most growing cells produce other products in addition to biomass which can be used to estimate the amount of growth. Carbon dioxide is a common metabolite which is produced from the oxidation of the carbon energy source, and is easily measured. Many anaerobic organisms do not have complete oxidation of the carbon energy source and consequently there is production of compounds such as lactic acid or ethanol. It is difficult to relate these products to growth directly, as their formation can be related to growth or non-growth. They are, however, usually better indicators of metabolic activity.

Production of hydrogen ions. When ammonium compounds are used as the hydrogen source, a corresponding hydrogen ion is formed as each mole of ammonium is consumed. In fermentation this is neutralized to maintain constant pH, and therefore the amount of alkali added can be related to growth. Alternatively, if nitrate is used as a nitrogen source, the amount of acid used for neutralization will be related to growth.

Growth yields of bacteria

For bacterial supplements to be economically justifiable in animal diets, they must be cost effective. This depends on producing the bacteria as cheaply as possible. Organisms vary in their ability to ferment substrate to bacterial biomass. It is, therefore, important to select an organism which is not only effective for a particular function, but also in the conversion of substrate to biomass. Predicting the potential yield of micro-organisms grown on complex media is inacccurate as no two fermentations are the same.

In bacteria, the breakdown of a defined quantity of substrate will give rise to the formation of a defined amount of new biomass. The yield of biomass produced from a substrate will depend on the amount of energy generated during breakdown and the amount of energy needed for the synthesis of new cell material together with the requirement for maintenance.

Molar growth yields

Cell yields of *B. subtilis* and *E. coli*, grown aerobically with a number of carbohydrates as carbon and energy sources, have been shown to be proportional to the amount of carbohydrate added. Similarly three strains of lactic acid bacteria (*Streptococcus faecalis*, *Leuconostoc mesenteroides* and *Lactobacillus delbrueckii*) were found to have a growth yield linearly related to carbohydrate consumption (DeMoss *et al.*,1951). Thus, the amount of growth is proportional to the substrate concentration and may be expressed as:

$$\text{Growth (g/ml)} = \text{Yield constant (K)} \times \text{Substrate concentration (g/ml)}.$$

The dry weight of bacteria and the substrate concentration are expressed in g/ml; K is the amount of dry weight in g formed during consumption of 1g of substrate. The value of K for a given substrate can be reproducibly determined and has a fixed value for every substrate.

Estimating the cell yields of lactic acid bacteria

For obligately fermentative micro-organisms such as *L. fermentum*, the biomass yield can be estimated if the yield of energy (as moles of ATP) is known per mole of the fermented carbon source (Bauchop and Elsden, 1960). It has been shown by many workers that the yield of biomass (i.e. cell mass produced in g) per mole of ATP derived from the carbon source (YATP) is approximately 10.5g/l for most micro-organisms (including *L. fermentum*) with a maximum variation 6 to 29g/l.

Homo-fermenters metabolise hexose through the Embden-Meyerhoff pathway (see *Figure* 9.12). However, hetero-fermenters cannot use this pathway since they lack the key enzyme fructose bisphosphate aldolase and consequently they metabolise hexose through the pentose phosphoketolase pathway (*Figure* 9.13) The major difference between the two pathways is their net ATP yield.

- Homo-fermentative (Emden-Meyerhoff) pathway (*Figure* 9.12)
  2 moles ATP per mole of hexose fermented
- Hetero-fermentative (Phosphoketolase) pathway (*Figure* 9.13)
  1 mole ATP per mole of hexose fermented

Homo-fermenters are preferable as they have higher yields per mole of substrate.

The current species of *Lactobacillus* and their classification in terms of fermentation are given in Table 9.8. This information is essential when determining fermentation conditions.

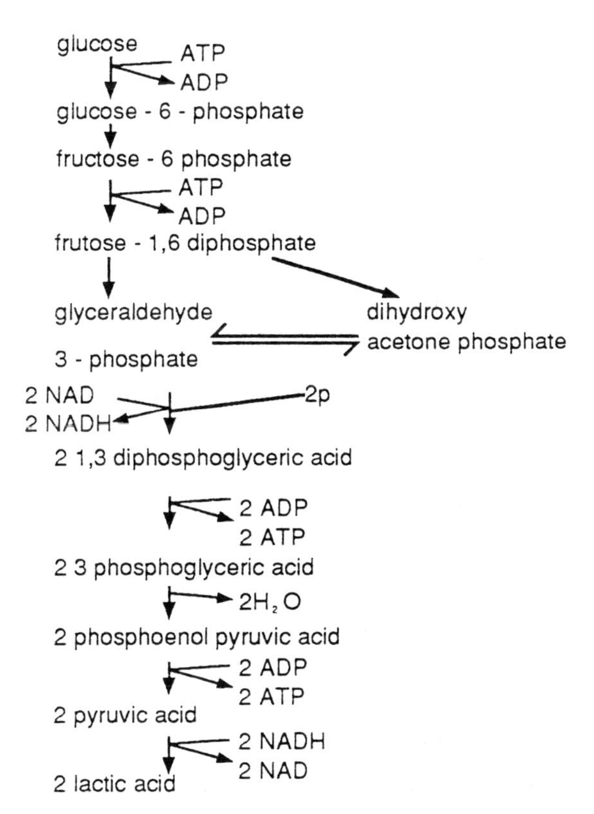

Figure 9.12 The Embden-Meyerhoff (homo-fermentative) pathway which yields 2 moles ATP per mole of hexose fermented.

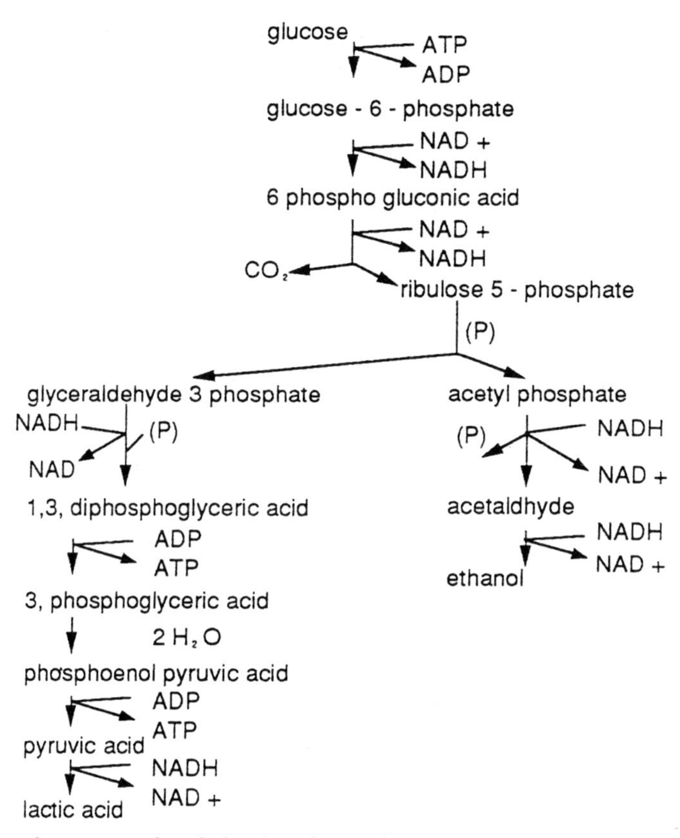

Figure 9.13 The pentose phosphoketolase (hetero-fermentative) pathway which yields 1 mole ATP per mole of hexose fermented.

Table 9.8 *Classification of* Lactobacilli *according to fermentation of hexose.*

Homo-fermentative Species	*Hetero-fermentative Species*
L. delbrueckii	*L. fermentum*
L. lactis	*L. cellobiosus*
L. casei	*L. brevis*
L. plantarum	*L. buchneri*
L. leichmanii	*L. viridescens*
L. jensenii	*L. coprophileus*
L. bulgaricus	*L. hilgardii*
L. helveticus	*L. trichodes*
L. acidophilus	
L. salvaricus	
L. xylosus	
L. curvatus	

A clear distinction can be made between *B. subtilis* and lactic acid bacteria (*Lactobacillus* spp. and *Streptococcus* spp.) as the latter are aeroduric anaerobes (can tolerate oxygen) while *B. subtilis* is an obligate aerobe (oxygen is essential for growth). The potential yield of biomass is much higher from *B. subtilis* than from any lactic acid bacterium. For this reason *B. subtilis* is often used in direct fed microbials.

The medium on which the organisms are growing greatly affects the yield of ATP for bacteria. In silage, for example, the sugars, cell walls, lactic acid, and VFAs all produce different ATP yields. It is important, therefore, that the medium is well defined if the growth and yield of bacteria are to be determined.

Determination of cell yield using *B. subtilis* as an example

The evaluation of aerobic growth is much more difficult than the evaluation of anaerobic growth yields. Under anaerobic conditions the ATP yield from the degradation of a substrate can be calculated exactly. However, under aerobic conditions it is difficult to supply sufficient oxygen to satisfy the maximum oxygen demand of the culture, even with the most advanced fermenters.

S. faecium, in common with other *Streptococci*, while being a homo-fermenter will produce fermentation products other than lactic acid (normally formate, ethanol and acetate) when grown under some conditions. The metabolic process for this switch is shown in *Figure 9.14*. Thus, under conditions of nutrient excess, the concentration of intermediates of hexose metabolism is high and pyruvate is converted quantitatively to lactate. However, under starvation conditions, some pyruvate is metabolised to ethanol and acetate.

The ability to use this metabloic switch allows more efficient use of the limiting amount of hexose, because ATP is generated during the conversion of pyruvate to acetate. *Streptococcus* species could therefore be grown more effectively on a fed-batch system (nutrients continually being added) rather than a batch culture system where nutrients are in excess.

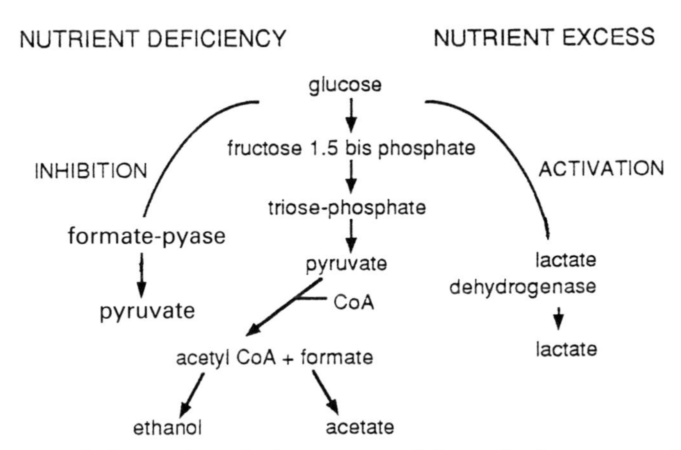

Figure 9.14 Metabolic switch in the fermentation of glucose by *Streptococcus faecium*, under different nutritional circumstances.

Bacterial supplements for the livestock feed industry

The livestock industry uses bacterial supplements produced by the fermentation methods discussed in many different ways. Haresign and Ewing (1989) reviewed the products commercially available and found that the three main physical forms were drenches/pastes, soluble powders and dry powders/granules (Table 9.9).

Haresign and Ewing (1989) also reviewed the main organisms used in products for sale on the UK market. Some live cultures contained only a single species of

Table 9.9 *Type of presentation of probiotics in relation to their method of application.*

Method of application	*Number of preparations*	
	Live culture	*Bacterial stimulant*
Readily constituted drench / paste / capsule for oral application	12	9
Soluble powder suitable for adding to milk replacers or for mixing with water and giving as a drench	20	2
Powder/granules for application as a top-dressing to the ration or for inclusion in pelleted feeds	19	5

*Includes some products which are presented in, or can be applied in, more than one form.

Table 9.10 *Number of single and multiple strain probiotic preparations containing live cultures of various strains of* Streptococcus faecium, Lactobacillus acidophilus *and/or* Bacillus subtilis *in 1989.*

Organism present	*Number of probiotic products*			
	Single species	*Two species*	*Three or more species*	*Not given*
S. faecium	15	19	8	–
L. acidophilus	1	10	9	–
B. subtilis	1	–	2	–
TOTAL	17	10*	10**	5

* All preparations contained both *S. faecium* and *L. acidophilus*

** 8 of these preparations contained *S. faecium* and *L. acidophilus* with either *B. subtilis* (2 preparations) or other organisms (6 preparations). The remaining 2 preparations contained either *L. acidophilus* or *S. faecium* together with other organisms.

(Haresign and Ewing, 1989)

micro-organisms, while others contained two or more species. The two most common organisms were *Streptococcus faecium* and *Lactobacillus acidophilus*, the strain varying between products. Thirty-six of the forty-two probiotics/bacterial supplements reviewed contained either one or both of these species (Table 9.10).

EU regulations

Regulations

Antibiotic manufacturers within the European Union are subject to its legislation. Directive 70/524 is central to their activities and the following are some of the key features.

Companies will be required to provide a detailed dossier of information regarding their products and claims, and nutrients in, to the Ministry of Agriculture Fisheries and Food, with the current proposed date of 1 January 1996. In the meantime, materials will have entered into the national list of the appropriate country by 1 Nobember 1994.

The data required in the dossier are summarised below:

1. Identity
   Name
   Type of additive
   Qualitative/quantitative
   Full composition (e.g. active ingredients, carriers, etc)
   Physical state
   Particle size
   Manufacturing process

- Specification of active substance
- Micro-organism name and taxonomic description
- If a fermentation product – formula, empirical and structural, molecular weight, both qualitative and quantitative.
- Degree of purity
- Relevant biological properties
- Manufacturing – purification processes and media used
- Biological properties of additive
- Stability of product
- Stability during feed processing
- Stability during storage
- Biological incompativilities or interactions
- Proposed use i.e.
  Species
  Administration method
  Dosage rate
  Withdrawal period
- Control method

2. Efficacy
   Results and experimental conditions

3. Microbiological studies of the additive:
   - Activity
   - Cross resistance
   - Effect on digestive tract
   - Resistance to antibiotics
   - Metabolism of product

4. Studies on excreted residues:
   - Nature and concentration of residue
   - Persistence

5. Acute toxicity

6. Mutagenicity

7. Subchronic toxicity

8. Chronic toxicity/carcinogenicity

9. Reproductive toxicity

Identification notes

Information required for identification has been detailed by Directive 93/113/EC for enzymes and micro organisms. Identification notes have to be submitted for the national lists.

1. Identity of the product

Trade name

The name or names under which the product is sold. An Identification Note should cover a single product.

Qualitative and quantitative composition

– **Active substance(s)**

– **Other components**

This should be a full product composition. Quantities should be in any easily recognised form, e.g. g/kg, % ,etc.

– **Impurities**

– **Undesirable substances**

Give maximum levels which will be specified by quality control procedures.

Name, address of manufacturer

This refers to the assembler of the product, as distinct from the person who holds the formulation or supplies raw materials.

Place of manufacture

Needs only be listed if different from registered place of business.

Name, address of person responsible for placing the product on the market, if not the manufacturer

The person responsible for placing the product on the market (and who submits the Identification Note) is the person who holds, owns or controls the formulation of the product, whether or not they actually manufacture that product.

2. Specification concerning the active substance

2.1 Micro-organisms

Name and taxonomic description. . .
Name and place of culture collection. . .
(If not registered and deposited, simply state this.)

State whether genetic manipulation has taken place
(If 'yes', brief details of the type of procedure should be included.)

Number of colony forming units
(Refer to the numbers in the active substance rather than the finished product.)

2.2 Enzymes

- **Name and taxonomic description**
  Name should follow International Union of Biochemistry and Molecular Biology recommendations.
- **State the biological origin**
  The notes on taxonomic description and culture collection given under micro-organisms above apply here as well.
- **State whether organism of origin has been genetically manipulated**
  Inflected brief details of procedures, if appropriate.
- **Relevant activities**
  Refer to activity in the specific enzyme rather than the finished product.

3. Properties of the product

Main effect
Information concerning effectiveness
Include a brief statement of the observed effects of any efficacy trials.

Justification for the presence of each active component
A brief indication of the function of each active component is required.

Other effects
Include any effects which will be claimed or advertised for the product. Note that submitting an Identification Note will not exempt a product from the controls of the Medicines Act if a medicinal claim (e.g. growth promotion) is made.

4. **Product Safety**

Available Information on safety

Include a brief description of any trials carried out or a brief summary of toxicological and relevant evidence to support the safety in use to humans, animals and the environment. Experimental details need not be presented at this stage, although full safety data will be required in the dossier. Reference to journals or other publications could be made, although copies of articles need not be included.

5. **Conditions of use**

This section covers the instructions for use that will be included with the product (e.g. target species,duration of use, level included in feed, etc). A clear indication of the point in the manufacturing process at which the product is used should be included (e.g. added to finished feed before presentation to animals, added during compounding). Product safety and risk prevention details should take relevant Health and Safety legislation into account.

6. **Technological information**

Stability of product

Should refer to recommendations on storage conditions, shelf-life and any special conditions required during feed manufacture. The description of the manufacturing process and quality control need only cover a basic outline.

7. **Control**

Methods of analysis

It is recognised that there are no statutory methods for enzymes and micro-organisms, although appropriate methods of analysis available or under development should be described.

8. **Attestation**

Suggested wording:

"I (name) (job title) of (company) certify that the information in this Identification Note is, to the best of my knowledge and belief, an accurate and complete description of the product (name of product)."

(Signature and date)

Further Information

Further information in the United Kingdom can be received from:

Chemical Safety Food Division
Ministry of Agriculture, Fisheries and Food
Room 531
Ergon House
c/o Nobel House
17 Smith Square
London SW1P 3JR

APPENDIX

Culture Collections

The following is a list of culture collections in various parts of the world.

Australia

CSIRO Division of Food Research (FRR)
P O Box 52
North Ryde
New South Wales 2113
Australia Tel. (612) 8878333
Telex AA 23407
Stocks. Yeasts, filamentous fungi, yeasts.

Brazil

Departamento Tecnologia Rural (ESALQ)
Avenue Padua Dias 11
Caixa Postal 9
Piracicaba
San Paulo
Brazil Tel 33 0011 R272
Stocks. Yeasts, bacteria filamentous fungi, yeasts.

Laboratories de Genetica de Micro-organisms
Departamento Microbiologia
University of Sao Paulo
Caixa Postal 4365
05508 Sao Paulo, SP
Brazil Tel. (011) 210–4311
Stocks. Yeasts.

Departamento de Micologia
Universidade Federal de Pernambuco
Rua. Prof. Morais Rego S/N
Cidade Universitaria
50.000 Recife, PE
Brazil Tel. (081) 271 3469
Stocks. Yeasts.

Laboratorios de Ecologia Microbiana e Taxonomia
Departamento de Microbiologia Geral
Instituto de Microbiologia
Universidade Federal do Rio de Janeiro
Ilha do Fundao/Cidade Universitaria
21944 Rio de Janeiro, RJ
Brazil Tel (021) 590 3093
Stocks. Yeasts.

Tropical Data Base (BDT) and Tropical Culture Collection (CCT)
Fundacao Tropical de Pesquisas e Tecnologia Andre Tosello
Rua Latino Coelho, 1301
13085 Campinas, SP
Brazil Tel. (0192) 427022
Stocks. Yeasts 100.

China

Centre for Microbiological Culture Collection (CGMCC)
China Committee for Culture Collections of Micro-organisms
(CCCCM)
Institute for Microbiology
P O Box 2714
Academia Sinica
Beijing
China Tel. 295614
Stock. Yeasts, bacteria, actinomycetes, yeasts, filamentous fungi, basidiomycetes.

Finland

VTT Collection of Industrial Micro-organisms (VTT)
VTT Biotechnical Laboratory
Tietotie 2
SF-02150 Espoo
Tel. 358 0 45615133 or 358 0 4565133
Telex 122972 VTTHA SF
Stocks. Yeasts, bacteria, yeasts, filamentous fungi.

Germany

Deutsche Sammlung von Mikroorganismen (DSM)
Mascheroder Weg 1b
D-3300 Braunschweig
Federal Republic of Germany Tel. 0531 6187–0
Telex 9 52 667 GEBIO-D
Electronic Mail TELECOM GOLD 75:DB10178
Stocks. Yeasts, bacteria, filamentous fungi, yeasts.

Bayerische Landesanstalt fur Weinbau und Gartenbau (LWG)
Residenzplatz 3
D-8700 Wurzburg
Federal Republic of Germany. Tel 0931/50701
Stocks. Yeasts.

Hungary

National Collection of Agricultural and Industrial Micro-organisms (NCAIM)
H-1118 Budapest
Somloi ut 14–16
Hungary Tel. 00 34 1 665466
Telex 226011
Stocks. Bacteria, filamentious fungi, yeasts.

India

National Collection of Industrial Micro-organisms (NCIM)
Division of Biochemical Sciences
National Chemical Laboratory
Pune 411008
India Tel 56451, 52, 53
Telex 0145.266
Stocks. Yeasts, bacteria, filamentous fungi, yeasts protozoa.

Italy

Yeast Collection of the Dipartimento di Biologia (DBV-PG)
Vegetale of the University of Perugia
c/o Dipartimento di Biologia Vegetale
Via 20 Giugno, 74
1–06100 Perugia
Italy Tel. 39 75 31766
Telex 216842 UMPG 11
Stocks. Yeasts and yeast-like fungi.

Japan

Institute of Applied Microbiology (IAM)
The University of Tokyo
1–1–1 Yayoi, Bunkyo-ku
Tokyo 113
Japan Tel 03 812 2111
Stocks. Yeasts, bacteria, filamentous fungi, yeasts, algae.

Japan Collection of Micro-organisms (JCM)
RIKEN
Wako-shi
Saitama 351–01
Japan Tel 0484 62 1111
Telex 2962818 (RIKENJ)
Cable RIKAGAKUINST
Stock. Yeasts, bacteria, actinomycetes, filamentous fungi, yeasts.

Institute for Fermentation, Osaka (IFO)
18–85 Juso-honmachi 2–chome
Yodogawa-Ku
Osaka 532
Japan Tel 06 302 7281
Stocks. Yeasts, bacteria, filamentous fungi, yeasts, phage, animal cell lines.

Netherlands

Centraalbureau voor Schimmelcultures (CBS)
(Yeast Division)
Julianalaan 67
2628 BC Delft
Netherlands Tel. 015–782394
Telex 38151 BUTUD NL
Stocks. Yeasts 4500.

Russia

All-Union Collection of Micro-organisms (VKM BKM)
Institute of Biochemistry and Physiology of Micro-organisms
Moscow Region
Pushchino 142292
Russia Tel. 3 05 26
Telex 3205887 OKA
Stocks. Yeasts 2000, actinomycetes, bacteria, filamentous fungi, yeasts.

Slovakia

Collection of Yeasts (CCY)
Institute of Chemistry
Slovak Academy of Sciences
Dubravska cesta 9
842 38 Bratislava
Slovakia Tel. 375000
Stocks. Yeasts, yeast-like organisms, algae.

Research Institute for Viticulture and Enology (RIVE)
Mataskova 25
83311 Bratislava
Slovakia
Tel Bratislava 46224/25/26
Stocks. Yeasts.

Spain

Coleccion Espanola de Cultivos Tipo (CECT)
Departmento de Microbiologia Facultad de Ciencias
Biologicas
Burjasot
Valencia
Spain Tel 96 3630011
Stocks. Yeasts, bacteria, filamentous fungi, yeasts.

Sweden

Culture Collection of the University of Goteborg (CCUG)
Department of Clinical Bacteriology
Guldhedsgatan 10
S41346 Goteborg
Sweden Tel 46.31 602016
(Electronic Mail TIXC AT SEGUC 21, DIALCOM 42: CDT0070, TELECOM GOLD
75:DB10070
Stocks. Yeasts, bacteria, filamentous fungi, yeasts.

Taiwan

Culture Collection and Research Centre (CCRC)
Food Industry Research and Development Institute
P O Box 246
Hsinchu 30038
Taiwan
Tel. (035) 223191
Stocks. Yeasts, bacteria, fungi.

Thailand

Thailand Institute of Scientific and Technological Research (TISTR)
196 Phahonyothin Road
Bangkok 10900
Thailand Tel (02) 5791121
Telex 21392 TISTR TH
Stocks. Yeasts 250, bacteria, filamentous fungi, yeasts, algae.

Turkey

Center for Culture Collections and Micro-organisms (KUKENS)
Istanbul Tip Fakultesi
Mikrobiyoloji Anabilim Dali
Temel Bilimler Binasi
Capa 34390
Istanbul, Turkey Tel. 5250904–5255504.

United Kingdom

National Collection of Type Cultures
Mycological Reference Laboratory
Central Public Health Laboratory
61 Colindale Avenue
London NW9 5HT Tel. 01 200 4000
Telex 8953942 DEFEND (G)
Stocks. Yeasts, filamentous fungi, yeasts.

National Collection of Yeast Cultures (NCYC)
Institute of Food Research, Norwich Laboratory
Colney Lane
Norwich
Norfolk NR4 7UA. Tel (0603) 56122
Telex 975453 (FRINOR G)
Electronic Mail TELECOM GOLD 75:db10013/db10151 0005
Stocks. Yeasts 2300.

Collection of the Commonwealth Mycological Institute (CMI)
Commonwealth Mycological Institute
Ferry Lane
Kew TW9 3AF
Surrey.

National Collection of Industrial and Marine Bacteria (NCIMB)
Torry Research Institute
135 Abbey Road
P O Box 31
ABERDEEN AB9 9DG.

United States of America (USA)

American Type Culture collection (ATCC)
12301 Parklawn Drive
Rockville
Maryland 20852
USA Tel (301) 881 2600
Telex 908 768 ATCC NORTH
Fax 301 231 5826
Stocks. Yeasts 8000 and most microbiological groups, plus animal and plant cultures.

Agricultural Research Service Culture Collection (NRRL)
Northern Regional Research Center
1815 North University Street
Peoria
Illinois 61604
USA Tel (309) 685 4011
Stocks. Yeasts.

Yeast Genetic Stock Center (YGSC)
Department of Biophysics and Medical Physics
University of California
Berkeley
California 94820
USA Tel (415) 642 0815
Stocks. Yeasts.

Bibliography

A.R.C. (1980). In : *The Nutrient Requirements of Ruminant Livestock*. Commonwealth Agricultural Bureaux, Farnham.

A.R.C. (1984). In: *The Nutrient Requirements of Ruminant Livestock* (Supp-1). Commonwealth Agricultural Bureaux, Farnham.

Abrams G.D., Bauer H. and Sprinz H. (1963). Influence of the normal flora on mucosal morphology and cellular renewal in the ileum. *Lab. Invest.*, 12:355–364.

Adams D.C., Galyean M.L., Keisling H.E., Wallace J.D and Finker M.D. (1981). *J. Anim. Sci.* 53:780–781.

Adetumbi M., Javor G.T., and Lau B.H.S. (1986). *Allium sativum* (garlic) inhibits lipid synthesis in *Candida albicans*. *Antimicrobial Agents and Chemotheraphy*. 30:499–501.

Adler, H.E. and Da Madssa, A.J. (1980). Effect of ingested *Lactobacilli* on *Salmonella infantis* and *Escherichia coli* and on intestinal flora, pasted vents, and chick growth. *Avian Dis.* 24: 868–878.

Alexander F. and Davies M.E. (1963). Production and fermentation of lactate by the bacteria in the alimentary canal of the horse and pig. *J. Comp. Pathol. Therap.* 73:1–8.

Allison M.J., Robinson I.M., Bucklin J.A. and Booth G.D. (1979). Comparison of bacterial populations of the pig cecum and colon based upon enumeration with specific energy sources. *Appl. Environ. Microbiol.* 37:1142–1151.

Alm L. (1983). The effect of *Lactobacillus acidophilus* administration upon the survival of *Salmonella* in randomly selected human carriers. *Prog. Food Nutr. Sci.* 7: 13–17.

Amer M., Taha M., and Tosson Z. (1980). The effect of aqueous garlic extract on the growth of dermatophytes. *International Journal of Dermatology*. 19:285–287.

Anderson G.W., Slinger S.J. and Pepper W.F. (1953). Bacterial cultures in the nutrition of poultry. 1. Effect of dietary bacterial cultures on the growth and cecal flora of chicks. *J. Nutr.* 50:35–36.

Annison G. and Chocht M. (1994). Plant polysaccharides – their physio-chemical properties and nutrional roles in monogastric animals. In: *Biotechnology in the Feed Industry* (Eds. T.P. Lyons and K.A. Jacques) pp. 51–66. Nottingham University Press, Nottingham.

Armstrong D.G. (1986). Gut-active growth promoters. In: *Control and Manipulation of Animal Growth* (Eds: P.J. Buttery, N.B. Haynes and D.B. Lindsay) 21–37. Butterworths.

Armstrong D. G. and Parker D. S. (1989). Gut-active growth enhancers – probiotics and antibiotics. *Anim. Prod.* 48 :631 (Abstr.).

Arp L.H., Jensen A.E. (1980). Piliation, hemagglutination, motility and generation time of *Escherichia coli* that are virulent or avirulent for turkeys. *Avian Dis.* 24:153.

Asano T. (1967). Inorganic ions in cecal content of gnotobiotic rats. *Proc. Soc. exp. Biol. Med.* 124: 424–430.

Attaix D. and Arnal M. (1987). Protein synthesis and growth in the gastrointestinal tract of the young pre-ruminant lamb. *Brit. J. Nutr.* 58: 159–169.

Attaix D, Aurousseau E., Manghebati A. and Arnal M. (1988). Contribution of liver, skin and skeletal muscle to whole – body protein synthesis in the young lamb. *Brit. J. Nutr.* 60: 77 – 94.

Baird D.M. (1977) Probiotics help boost feed efficiency. *Feedstuffs.* 49, 11–12.

Baker N.R. and Marcus H. (1982). Adherence of clinical isolates of *Pseudomonas aeruginosa* to hamster tracheal epithelium in vitro. *Curr. Microbiol.* 7:35–40.

Balger G, Hoerstke M., Dirksen G., Sietz A., Sailer J., and Mayr A. (1981). *Zentralblatt fur Veterinarmedzin B*, 28; 759–796

Barber R.S., Braude R., Mitchell K.G., Rook J.A.F. and Rowell J.G. (1957). Further studies on antibiotic and copper supplements for fattening pigs. *Brit. J. Nutr.* 11: 70 – 79.

Bare L.N. and Wiseman R.F. (1964). Delayed appearance of *Lactobacilli* in the intestines of chicks reared in a "new" environment. *Appl. Microbial.* 12: 457–459.

Barnes E.M, Impey C.S. and Cooper D.M. (1980a). Competitive exclusion of *Salmonella* from the newly hatched chick. *Vet. Rec.* 106: 61.

Barnes E.M, Impey C.S. and Cooper D.M. (1980b). Manipulation of the crop and intestinal flora of the newly hatched chick. *Am. J. Clin. Nutr.* 33: 2426–2433.

Barnes E.M., Mead G.C., Barnum D.A. . and Harry E.G. (1972). The intestinal flora of the chicken in the period 2 to 6 weeks of age, with particular reference to the anaerobic bacteria. Competitive exclusion of salmonellas from the newly hatched chick. *Br. Poult. Sci.* 12:311–326.

Barnes E.M. and Impey C.S. (1974). The occurrence and properties of uric acid decomposing anaerobic bacteria in the avian cecum. *J. Appl. Bact.* 37:393

Barnes H.J. (1987) *Escherichia coli-* problems in poultry production. In: *Alltech's Third Annual* Symposium. Biotechnology in the Feed Industry (Ed T.P. Lyons). Lexington, Kentucky, USA. Alltech Technical Publications, Nicholasville, Kentucky.

Barrow P.A., Fuller R. and Newport M.J. (1977). Changes in the micro-flora and physiology of the anterior tract of pigs weaned at 2 days with special reference to the pathogenesis of diarrhoea. *Infect. Immun.* 18, 586–595.

Barrow P.A., Brooker, B.E., Fuller R. and Newport M.J. (1980). The attachment of bacteria to the gastric epithelium of the pig and its importance in the micro-ecology of the intestine. *J. Appl. Bact.* 48: 147–154.

Bauchop T. and Elsden S.R.(1960). The growth of micro-organisms in relation to their energy supply. *J. Gen. Micr.* 23: 457–469.

Bechman T.J., Chambers J.V and Cunningham M.D, (1977). Influence of *Lactobacillus acidophilus* on performance of young dairy calves. *J. Dairy Sci.* (Suppl. 1): 74.

Beck T. (1972). The microbiology of silage fermentation. *Fermentation of Silage – A Review.* National Feed Ingredient Association., Des Moines, Iowa (Ed. M.E. McLullogh) 332.

Beers S. and Jongbloed A.W. (1992). Effect of supplementary *Aspergillus niger* phytase in diets for piglets on their performance and apparent digestibility of phosphorus. *Anim. Prod.* 55:425–430.

Beers S. and Koorn T. (1990). *Report IVVO No.* 223, Lelystad. p.24.

Benno Y. and Mitsuoka T. (1986). Development of the intestinal micro-flora in humans and animals. *Bifidobacteria Microflora* 5: 13–25.

Bergey L.L. (1986) *Manual of Determinative Bacteriology.* Williams and Wilkins, Baltimore, USA.

Berridge N.J. (1949). Preparation of the antibiotic Nisin. *Biochem.* J. 45:486–493.

Bhattachary P.R. and Majunder M.K. (1983). Survival of orally administered isolated intestinal *Lactobacillus acidophilus* in different parts of gastro-intestinal tract of mice. *J. Bio. Sci.* 5: 97–105.

Biochemical basis for microbial antagonism in the intestine. In: *Natural Antimicrobial Systems.* (Eds. Gould. G.W., Rhodes-Roberts, X.E., Charnley, A.K., Cooper, R.M. and Board, R.G). pp. 29, Bath University Press.

Blaxter K.L. and Wood W.A. (1953). Some observations on the biochemical and physiological events associated with diarrhoea in calves. *Vet. Rec* 65: 889–892.

Blecha F., Pollman D.S. Nichols D.A. (1983). Weaning pigs at an early age decreases cellular immunity *J. Anim. Sci.* 56:396–400.

Blecha F., Boyles S.L., Riley G.G., (1984). Shipping suppresses lymphocyte blastogenic responses in Angus and Braham X Angus feeder calves *J. Anim. Sci.* 59:576–583.

Bohnhoff M.B., Drake L. and Miller C.P. (1954). Effect of streptomycin on susceptibility of intestinal tract to experimental *Salmonella* infection. *Proc. Soc. Exptl. Biol. and Med.* 86:132–137.

Bohnoff M., Miller C.P. and Martin W.R. (1964). Resistance of the mouse's intestinal tract to experimental Salmonella infection. I Factors which interfere with the initiation of infection by oral inoculation. *J. Expt. Med.* 20:805–816.

Boldt D.H. and Banwell J.G. (1985). Binding of isolectins from raw kidney bean *(Phaseolus Vulgaris)* to purified rat brush-border membranes. *Biochema et biophysica Acta* 843, 230–237.

Borland E.D. (1975). *Salmonella* infections. *Poult. Vet. Rec.*, 97; 406–608.

Bousefield I.J., Smith G.L. and Trueman R.W. (1973). The use of semi-automatic pipettes in the viable counting of bacteria. *J. Appl. Bact.* 36: 297–299.

Briggs M. (1953). An improved medium for *Lactobacilli J. Dairy Res.* 20:36–40.

Brown L.D., Jacobson D.R., Everett J.P., Seath D.M. and Rust J.W. (1960). Urea utilization by young dairy calves as affected by chlortetracycline supplementation. *J. Dairy Sci.* 43: 1313.

Brownlee A. and Moss W. (1961). The influence of diet on *Lactobacilli* in the stomach of the rat. *J. Pathol. Bacteriol.* 82: 513–516.

Bruce B.B., Gilliland S.E., Bush L.J. and Staley T.E. (1979). Influence of feeding cells of *Lactobacillus acidophilus* on the faecal flora of young dairy calves. *Oklahoma Anim. Sci.*

Bryant M.P., Small N. Bouma C. and Robinson I. (1958). Studies on the composition of the ruminal flora and fauna of young calves. *J. Dairy Sci.* 41: 1747–1767.

Bryant M.P. and Small N. (1960). Observations on the ruminal microorganisms of isolated and inoculated calves. *J. Dairy Sci.* 41: 1747–1767.

Buchanan X.P. and Gibbons N.E. (Eds.) (1974). *Bergey's Manual of Determinative Bacteriology.* (8th ed.) Williams and Wilkins Co., Baltimore, MD.

Buckland B.C. (1988) Personal Communication.

Bull A.T., Ellwood D.C. and Ratledge C. (1979). The changing scene in microbial technology. *Soc. Gen. Microbiology Symposium* 29:1–28.

Buraczewska L. (1988). *Proceedings of 4th International Seminar on Digestive Physiology in Pigs.* p. 66. Warsaw, Poland.

Burnett G.S. and Neil E.L. 1977. A note on the effect of probioticum feed additive on the live-weight gain, feed conversion and carcass quality of bacon pigs. *Anim. Prod.* 25: 95–98.

Bush L.J., Allen R.S. and Jacobson N.L. (1959). Effect of chlortetracycline on nutrient utilization by dairy calves. *J.Dairy Sci.* 42:671.

Calcott P.H. (1981). The construction and operation of continuous cultures. In. *Continuous Culture of Cells,* pp 13–26 vol 1, (Ed. Calcott P.H.) CRC Press, Boca Raton.

Carlstedt-Duke B., Alm L., Hoverstad T., Midtvedt A.C., Norin K.E., Steinbak M. and Midtvedt T. (1987). Influence of clindamycin, administered together with or without *Lactobacilli,* upon intestinal ecology in rats. *FEMS Microbiol. Ecol.* 45:251–259.

Catton R. (1985). Studies of factors affecting the nutritional value of feedstuffs in complete diets for ruminants. Ph.D. Thesis, University of Wales.

Chan R., Lian C.J., Costerton J.W. and Acres S.D. (1982b). The use of specific antibodies to demonstrate the glycocalyx and spatial relationships of Kk99–, F41–, enterotoxigenic strain of *Escherichia coli* colonising the ileum of colostrum-deprived calves. *Can. J. Comp. Med.* 47:150–156.

Chapman J.D. (1988). Probiotics, acidifiers and yeast culture: A place for natural additives in pig and poultry production. In *Biotechnology in the Feed Industry,* (Ed. T.P. Lyons), Alltech Technical Publications, Kentucky. pp. 45–54.

Cheney C.P., Schad P.A. Formal S.B. and Boedeker E.C. (1980). Species specificity of in vitro *Escherichi coli* adherence to host intestinal cell membranes and its correlation with in vivo colonisation and infectivity. *Infect. Immun.* 28:1019.

Cheng K.J., Akin D.E. and Costerton J.W. (1977). Rumen bacteria: interaction with particulate dietary components and response to dietary variation. *Fed. Proc.* 36:193–197.

Cheng K.J., Irvin R.T. and Costerton J.W. (1981). Autochthonous and pathogenic colonisation of animal tissues by bacteria *Can. J Microbiol.* 27: 461–490.

Cheplin H.A., Post C.D. and Wiseman J.R. (1923). *Bacillus acidophilus* milk and its therapeutic effects. *Boston Med and Sci J.* 189:405.

Cheplin H.A. and Rettger L.F. (1922). The therapeutic application of *Lactobacillus acidophilus*. *Abs. Bact.* 6:60.

Chesson A. (1993) Probiotics and other intestinal mediators. In: *Recent Advances in Animal Nutrition.* (Eds P.C. Garnsworthy and D.J.A. Cole) Nottingham University Press. Nottingham, England.

Chesson A. (1993). Probiotics and other intestinal mediators. In: *Principles of Pig Science.* (Eds D.J.A. Cole, J.Wiseman and M.A. Varley) Nottingham University Press, Nottingham, England.

Chidlow J.W. (1979). The role of maternal immunity in neonatal protection against enteric disease. *Pig Vet. Soc. Proc.* 4:31–40.

Chopra S.L., Blackwood A.C. and Dale D.G. (1963). Intestinal micro-flora associated with enteritis of early-weaned pigs. *Can. J. Comp. Med. Vet. Sci.* 27: 291–294.

Clegg S. and Gerlach G.F. (1987). Enterobacterial fimbria *J. Bacteriol.* 169: 934–938.

Cline T.R., Forsyth D. and Plumlee M.P. (1976). Probios for starter and grower pigs. *Purdue Swine Day Rep.* : 53–59.

Close W. (1992). *Enzymes in Pig and Poultry Production.* R&H Hall Technical Bulletin No. 3

Dublin.Hazlewood G.P., Gilbert H.J., Rixon J.E., Sharp R.S., and O'Donnell A.G. (1993). The use of genetically engineered *Lactobacillus plantarum* in the ensilage process. *Anim. Prod.* 56:458 (Abstr.).

Coates M.E., Davies M.K. and Kon S.K. (1955). The effect of antibiotics on the intestine of the chick. *Br. J. Nutr.* 9:110.

Coates M.E., Fuller R., Harrison G.F. Lev M. and Suffolk S.F. (1963). A comparison of the growth of chicks in the Gustafson germfree apparatus and in a conventional environment with and without dietary supplements of penicillin. *Br. J. Nut.r* 17, 141–150

Coates M.E. (1976) In: *Digestion in the Fowl.* pp. 179–191. (Ed. Boorman K.N. and Freeman C.P). 1st ed. Br. Poultry Sci. Ltd, Edinburgh.

Coates M.E. (1980). The gut micro-flora and growth. In: *Growth in Animals,* 175–188. (Ed. Lawrence T.L.J.), Butterworths, London.

Cole C.B. and Fuller R. (1984). A note on the effect of host specific fermented milk on the coliform population of the neonatal rat gut. *J. Appl. Bact.* 56: 495–498.

Cole C.B., Fuller R. and Newport, M.J. (1987). The effect of diluted yoghurt on the gut microbiology and growth of piglets. *Food Microbiol.* 4: 83–85.

Cole C.B., Anderson P.H., Philips S. M. Fuller R. and Hewitt D. (1984). The effect of yoghurt on the growth, lactose-utilising gut organisms and B-glucorinidase activity of caecal contents of a lactose-fed, lactose-deficient aninmal. *Food Microbiol.* 1: 217–222.

Cole D.J.A. (1991) The role of the nutritionist in designing feeds for the future. In: *Biotechnology in the Feed Industry* (Ed. T.P. Lyons) 1–20. Alltech Technical Publications, Nicholasville, Kentucky.

Cole D.J.A., Beal R.M. and Luscombe J.R. (1968). The effect on performance and bacterial flora of lactic acid, propionic acid, calcium propionate and calcium acrylate in the drinking water of weaned pigs. *Vet. Rec.* 83, 459–464.

Cole D.J.A., Beal R.M. and Luscombe J.R. (1970). Further studies on the control of *E.coli* in weaned pigs by chemical supplementation of the feed. *Vet. Rec*, 86: 400–404.

Cole D.J.A., Wiseman J., Davies J., Partridge G.C. and Reeve A. (1992). The effect of dietary electrolyte balance on the performance of growing pigs. *Anim. Prod.* 54: 485.

Collington G.K., Parker D.S., Ellis M. and Armstrong D.G. (1988). The influence of probios or tylosin on growth of pigs and development of the gastro-intestinal tract. *Anim. Prod.* 46: 521 (Abstr).

Collins E.B. and Aramaki K. (1980). Production of hydrogen peroxide by *Lactobacillus acidophilus. J. Dairy Sci.* 63: 353–357.

Collins C.H. and Lyne P.M. (1976). *Microbiological Methods* (4th ed.) Butterworths: London – Public Health Laboratory Service.

Contrepois M.G. and Girardeau J.P. (1985). Additive protective effects of colostral antipili antibodies in calves experimentally infected with enterotoxigenic *Escherichia coli. Infect. Immun.* 50: 947–949.

Conway P.L., Gorbach S.L. and Goldin B.R. (1987). Survival of lactic acid bacteria in the human stomach and adhesion to intestinal cells. *J. Dairy Sci. 70:1–12.*

Conway P.L. and Kjelleberg S. (1989). Protein-mediated adhesion of *Lactobacillus fermentum* strain 737 to mouse stomach squamous epithelium. *J. Gen. Microbiol.* 135:1175–1186.

Cook R.L. Harris R.J. and Reid G. (1988). Effect of culture media and growth phase on the morphology of *Lactobacilli* and on their ability to adhere to epithelial cells. *Curr. Microbiol.* 17: 159–166.

Cooke R.H. and Bird F.H. (1973). Duodenal villus area and epithelial cellular migration in conventional and germ-free chicks. *Poult. Sci.* 52:2276.

Coppoolse J., van Hurren J., Huisman J., Janssen W.M.M.A., Jongbloed A.W., Lenis N.P. and Simons P.C.M, (1990). *De nitscheiding van stikstof, fodfor en Kalium door Landbouwhuissieren. Nu en Morgen*, IVVO. COVP. ILON-TNO, The Netherlands.

Costerton J.W., Marrie T.J. and Cheng K.J. (1985). Phenomena of bacterial adhesion. In: *Bacterial Adhesion.* (Ed. Savage D.C. and Fletcher M.) 3–43. Plenum Press, New York.

Cowan G.L., Davis L.W., Duncan M.S. and Trammell J.H. (1978). Dose relationship in post-weaned pigs fed a non-viable, *Lactobacillus* fermentation product. *Proc. National Amer. Soc. Anim. Sci.* 298 (Abstr).

Cox C.P. and Briggs M. (1954). Experiments on growth media for *Lactobacilli. J.appl. Bact.* 17:18–20.

Crawford J.S. (1979). Probiotics in animal nutrition. *Proc. Arkan. Nutr.Conf.* 45–55.

Crawford J.S., Carver L., Berger J. and Dana, G. (1980). Effects of feeding living non-freeze-dried *Lactobacillus acidophilus* culture on performance of incoming feedlot steers. *Proc. Amer. Soc. Anim. Sci. West. Sect.* 31: 210–212.

Cummings J.H. and Macfarlane G.T. (1991). A review: The control and consequences of bacterial fermentation in the human colon. *J. Appl. Bacteriol.* 70:443–459.

Dahyia R.S. and Speck M.L. (1968). Hydrogen peroxide formation by *Lactobacilli* and its effect on *Staphylococcus aureus*. *J. Dairy Sci.* 51: 1568–1572.

Damron B.K., Wilson H.R., Voitle, R.A. and Harms R.H. (1981). A mixed *Lactobacillus* culture in the diet of Broad Breasted Large White turkey hens. *Poultry Sci.* 60: 1350–1351.

Damron P. (1987). Effectiveness of the lactic acid bacteria *Streptococcus faecium* M-74 in feed mixtures for early-weaned piglets. *Nutr. Abst. Rev.* (Series B) 57: 364.

Danielson A.D., Peo E.R., Shahani K.M., Lewis A.J., Whalen P.J. and Amer M.A. (1989). Anticholesteremic property of *Lactobacillus acidophilus* yoghurt fed to mature birds. *J. Anim. Sci.*, 67: 966–973.

Davidson J.N. and Hirsch D.C. (1976). Bacterial competition as a means of preventing neonatal diarrhoea in pigs. *Infect. Immun.* 13: 1773–1774.

Davis A.V. and Woodward R.S. (1979). Response of calves fed a *Lactobacillus* fermentation product. *J. Dairy Sci.* 62 (Suppl. 1): 104.

Davis J.G. (1960). The *Lactobacilli* 1. *Proc. Ind. Microbiol* 2 1–26.

Davis P.J, Barratt M.E.J, Morgan M. and Parry S.H, (1986). In: *Proc. Georgia Coccidiosis Conference* pp. 618–633. (Eds. McDougald L.R., Joyner L.P. and Long P.L). University of Georgia Press, USA.

Dawson K.A. (1990) Designing the yeast culture of tomorrow – mode of action of yeast outline for ruminants and non-ruminants. In: *Biotechnology in the Feed Industry* (Ed. T.P. Lyons), pp. 59–75. Alltech Technical Publications, Nicholasville.

De Man J.C., Rogosa M. and Sharpe M.E. 1960. A medium for the cultivation of *Lactobacilli*. *J. Appl. Bact*, 23: 130–135.

Deibel R.H. and Seeley H.W. (1974). Gram positive cocci. In *Bergey's Manual of Determinative Bacteriology*, pp. 490–491. Waverly Press, USA.

Dellaglio F., Bottazzi V. and Trovatelli L.D. (1973). Deoxyribonucleic acid homology and base compostion in some thermophilic *Lactobacilli*. *J. Gen. Microbiol* 74: 289–297.

Dellaglio F., Bottazzi V. and Vescovo M. (1975). Deoxyribonucleic acid homology among *Lactobacillus* species of the subgenus *streptobacterium*. *Int. J. Syst. Bacteriol*, 25: 160–172.

Delluva M., Markley K. and Davies R.F. (1968). The absence of gastric urease in germ-free animals. *Biochimica et Biophysica Act.*, 151: 646–650.

Dew R.K. and Thomas 0.0. (1981). *Lactobacillus* fermentation product for post-weaned calves. *J. Anim. Sci.* 53 (Suppl. 1): 483.

Dierick N.A. and Decuypere J.A. (1994). Enzymes and growth in pigs. In: *Principles of Pig Science*. (Ed. D.J.A. Cole, J. Wiseman and M.A. Varley). pp. 169–195, Nottingham University Press, Nottingham.

Dilworth B.C. and Day E.J. (1978). *Lactobacillus* Cultures in broiler diets. *Poultry Sci.* 57: 1101 (Abstr.).

Donaldson R.M. (1964). Normal bacteria populations of the intestine and their relationship to intestinal function. *New England J. Med.* 270: 994.

Dubos R., Schaedler R.W., Costello R. and Hoet P. (1965). Indigenous, normal and autochthonous flora of the gastro-intestinal tract. *J. Exp. Med.* 122:67–76.

Dubos R.J. and Schaedler R.W. (1962), Effect of diet on fecal bacterial flora of mice and on their resistance to infection. *J.Exp Med.* 115: 1161–1172.

Dubos R.J. and Schaedler R.W. (1960). The effect of the intestinal flora on the growth rate of mice, and on their susceptibility to experimental infections. *J. Exptl. Med.*, 111:407–417.

Ducluzeau R. (1985). Implantation and development of the gut flora in the newbom piglet. *Pig News and Information.* 6: 415–418.

Ducluzeau R., Raibaud P., Dubos F., Clara A. and Lhuillery C. (1981). Permanent effect of some dietary regimens on the establishment of two *Clostridium* strains in the digestive tract of gnotobiotic mice. *Anim. J. Clin. Nutr.* 34: 520–526.

Ducluzeau R., Dubos F., Martinet L. and Raibaud P. (1975). Digestive tract micro-flora in healthy and diarrheic young hares born in captivity. Effect of intake of different antibiotics. *Ann. Biol. Anim. Biochim. Biophys.* 15: 529–535.

Ducluzeau R. (1983). Implantation and development of the gut flora in the newborn animal. *Ann. Rech. Vet.* 14: 354–359.

Dudgeon L.S. (1926). Study of intestinal flora under normal and abnormal conditions. *J.Hyg.* 25: 119 14.

Easter R.A. (1988). Acidification of diets for pigs. In: *Recent Advances in Animal Nutrition.* (1988). (Eds. Haresign W. and Cole D.J.A.). Butterworths, London.

Edmunds B.K., Buttery P.J. and Fisher C. (1980). Protein and energy metabolism in the growing pig. In: *Energy Metabolism* (Ed. Mount L.E.) pp. 129–133. Butterworths, London.

Edwards I.E., Mutsvangwa T., Topps J.H. and Paterson G.F.M. (1990). The effect of yeast culture (Yea-Sacc) on patterns of rumen fermentation and growth performance of intensively fed bulls. *Anim. Prod.* 50, 579 (Abstr).

Ellen R.P. and Gibbons R.J. (1972). M-protein associated adherence of *Streptococcus pyogenes* to epithelial surfaces: pre-requisite for virulence. *Inf. and Immun,* 5: 826–830.

Ellinger D. K., Muller L.D. and Glantz P.J. (1980). Influence of feeding fermented colostrum and *Lactobacillus acidophilus* on faecal flora of dairy calves. *J. Dairy Sci.* 63: 478–482.

Erasmus L.J., Botha P.M. and Kistner A. (1992). Effect on yeast culture supplementation production, rumen fermentation and duodenal nitrogen flow in dairy cows. *J. Dairy Sci.* 75: 3056–3065.

Erasmus L.J. (1991). The importance of the duodenal amino acid profile for dairy cows ad the impact of changes in these profiles of following the use of Yea-Sacc [1026]. *Feed Compounder.* August 1991.

Evans P.A., Newby T.J., Stokes C.R., Patel D. and Bourne F.J. (1980). Antibody response of the lactating sow to oral iminisation with *Escherichia coli. Scand. J. Immunol.* 11: 419–430.

Ewing W.N. and Cole D.J.A. (1988). In: *Probiotics*: Theory and Applications. (Eds. B.A. Stark and J.M. Wilkinson). Chalcombe Publications, Wye College, England. pp. 39–45. In Porter W.L. (1989) Practical applications of probiotics in livestock production.

Ewing W.N. and Haresign W. (1989). *Probiotics.* Chalcombe Publications, Wye, England.

Eyssen H., Swaelen E., Kowszk-Gindifer Z. and Parmenteer G. (1965). Nucleotide requirements of *Lactobacillus acidophilus* variants isolated from the crops of chicks. *Antonie van Leeuwenhoek.* 31: 241.

Eyssen H. and Desomer P. (1967). Effects of *Streptococcus faecalis* and a filterable agent on growth and nutrient absorption in gnotobiotic chicks. *Poultry Sci.* 46: 323–333.

Eyssen J., De Prins V. and De Somer P. (1962). The growth promoting effect of virginiamycin and its influence on the crop flora in chickens. *Poultry Sci,* 41: 227.

Fallon R. (1986). Calf – the European experience. In: *Second Ann. Biotech. Symp.* (Ed: T.P. Lyons) Alltech, Inc. Nicholasville, Kentucky.

Fallon R.J. and Harte F.J. (1987). *Irish Grassland and Animal Production Journal.* 21, 156 (Abst).

Fauconneau G. and Michel M.C. (1970). *Mammalian Protein Metabolism.* Vol. 4, 5, 497–506. (Ed H.N. Munro). Academic Press, New York.

Feingold D.S. (1963). Antimicrobial chemotherapeuric agents: the nature of their action and selective toxicity. *New Eng. J. Med.*, 269. 900–907, 957–964.

Fethiere R. and Miles R.D. (1988). Intestinal tract weight of chicks fed on antibiotic and probiotic. *Nutr. Abstr. Rev.* (Series B) 58:686.

Fevrier C., Ducluzeau R. and Vassal L. (1979). Feeding fattening pigs by Lehmann's method with skimmed milk or yoghurt with or without live bacteria. *Journ de la Recherche Porcine en France.* 291–298.

Finlay B.B., Heffron F. and Falkow S. (1989). Epithelial cell surfaces induce *Salmonella* proteins required for bacterial adherence and invasion. *Science.* 940–943.

Floch M.H., Gershengoren W., Diamond S. and Hersh T. (1970). Cholic acid inhibition of intestinal bacteria. *Amer. J. Clin. Nutr,* 23:8–10.

Floch M.H., Binder H.J., Filburn B. and Gershengoren M.S. (1972). The effect of bile acids on intestinal micro-flora. *Amer. J. Clin. Nutr.* 25:1418–1421.

Forbes M. and Park J.T. (1959). Growth of germ-free and conventional dietary penicillin and bacterial environment. *J. Nutr,* 67: 69.

Fox J.G. and Lee A. (1989). Gastric campylobacter-like organisms: their role in gastric disease of laboratory animals. *Lab. Animal Sci.* 39:543–553.

Francis C., Tanky D., Arafa A. and Harm R. (1978) Inter-relationship of *Lactobacillus* and Zinc bacitracin in the diets of turkey poults. *Poultry Sci.* 57:1687–1689.

Freter R. (1956). Experimental enteric *Shigella* and Vibrio infection in mice and guinea pigs. *J. of Exp. Medicine.* 104: 411–418.

Freter R. (1983). Mechanisms that control the micro-flora in the large intestine. In *Human Intestinal Micro-flora in Health and Disease.* (Ed Hentges D.J.), Academic Press, New York. pp. 33–54.

Friend B.A. and Shahani K.M. (1984). Antitumor properties of *Lactobacilli* and dairy products fermented by *Lactobacilli. J. Food Prot.* 47:717–721.

Frobisher M., Hinsdill R.D., Crabtree K.T. and Goodheart C.R. (1974). The *Cocci: Streptococci.* In *Fundamentals of Microbiology,* pp. 561–566. Philadelphia; WB. Saunders company.

Fromtling R. and Bulmer G.S. (1978). *In vitro* effect of aqueous extract of garlic *(Allium sativum)* on the growth and viability of *Cryptococcus neoformis. Mycologia.* 70, 397–405.

Frumholtz P.P., Newbold C.J. and Wallace R.J. (1989) Influence of *Aspergillus oryzae* fermentation extract on the fermentation of basal ration in the rumen simulation technique (Rusitec). *J. Agric. Sci., Cambridge.* 113 169–172.

Fuller R. (1973). Ecological studies on the *Lactobacillus* flora associated with the crop epithelium of the fowl. *J. Appl. Bact.* 36: 131–139.

Fuller R., Coates M.E. and Harrison GF (1979). The influence of specific bacteria and a filterable agent on the growth of gnotobiotic chicks. *J. Appl. Bacteriol.* 46: 335–342.

Fuller R., Houghton S.B. and Brooker B.E. (1981). Attachment of *Streptococcus faecium* to the duodenal epithelium of the chicken and its importance in colonisation of the small intestine. *Appl. Environ. Microbiol.* 41: 1433–1441.

Fuller R. (1984). Microbial activity in the alimentary tract of bird. *Proc. Nutr. Soc.* 43: 55–61.

Fuller R., Barrow P.A. and Brooker B.E. 1978. Bacteria associated with the gastric epithelium of neonatal pigs. *Appl. Environ. Microbial.* 35: 582–591.

Fuller R. (1989) Probiotics in man and animals. *J. Appl. Bact.*, 66:365–378.

Fuller R. and Turvey A. (1971). Bacteria associated with the intestinal wall of the fowl *(Gallus domesticus)*. *J. Appl. Bact.* 34(3):617–611.

Fuller R., Barrow P.A. and B.E. Brooker (1978). Bacteria associated with gastric epithelium of neonatal pigs. *Appl. Environ. Microbiol.* 35:582–591.

Fuller R. (1977). The importance of *Lactobacilli* in maintaining normal microbial balance in the crop. *Br. Poult Sci.* 18: 85–94.

Fuller R. (1978). Epithelial attachment and other factors controlling the colonization of the intestine of the gnotobiotic chicken by *Lactobacilli*. *J. Anim. Bact.* 45: 389–395.

Fuller R. (1982). Development and dynamics of the aerobic gut flora in gnotobiotic and conventional animals. *Adv.Vet. Med.* 33: 7–15.

Fuller R. and Brooker B.E. (1974). *Lactobacilli* which attach to the crop epithelium of the fowl. *Am. J. Clin. Nutri.* 27:1305–1312.

Fuller R. and Brooker B.E. (1980). The attachment of bacteria to squamous epithelial cells. In: *Microbial Adhesion to Surfaces*, pp. 495–507. Ellis Horwood Ltd., Chichester.

Fuller R. and Cole C.B. (1988) The scientific basis of the probiotic concept. In *Probiotics – Theory and Applications* pp. 1–14. (Eds B.A. Stark and Wilkinson J.M.) Chalcombe Publications, Marlow.

Fuller R. and Turvey A. (1971). Bacteria associated with the intestinal wall of the fowl. *J. App. Bacteriol.* 34: 617–622.

Furuse M. and Yokota H. (1985) Effect of the gut micro-flora on chick growth and utilisation of protein and energy at different concentrations of dietary protein. *Br. Poult. Sci.* 26: 97–104.

Gaastra W. and De Graaf F.K. (1982). Host-specific fimbrial adhesions of non-infective enterotoxigenic *Escherichia coli* strains. *Microbiol Rev.* 46:,129–301.

Galjaard H., Meer-Fieggen W. van der and Giesen J. (1972). Feedback control by functional villus cells on cell proliferation and maturation in intestinal epithelium. *Expl. Cell Res*, 73: 197–207.

Gandhi D.N. and Nambudripad U.K.N. (1978) Implantation of *Lactobacillus acidophilus* in the intestines of adults suffering from gastro intestinal disorders. In Vol E. 972, XX. *International Dairy Congress*. E. 972.

Gasser F. and Mandel M. (1968). Deoxyribonucleic acid base composition of the genus *Lactobacillus*. *J. Bact.* 96:580–588.

Gibbons R.J. (1984). Adherence of the oral flora. In *Attachment of Organisms to the Gut Mucosa*, Vol.l. (Ed. E.C. Boedeker). pp. 11 – 20 CRC Press, Florida.

Gibbons R.J. and Van Houte J. (1971) Selective bacteria adherence to oral epithelial surfaces and its role as an ecological determinant *Infect. Immunity*. 3:567–573.

Gibson G.R., Macfarlane G.T., Cummings J.H. (1993). Sulphate reducing bacteria and hydrogen metabolism in the human large intestine. Gut. 34:437–439.

Gibson T., Stirling A.C., Keddle R.M. and Rosenberger R.F. (1958). Bacteriological changes in silage made at controlled temperature. *J. Gen. Micr.* 19:112–129.

Gibson T., Stirling A.C., Keddie R.M. and Rosenberger R.F. (1961). Bacteriological changes in silage as affected by laceration of the fresh grass. *J. Appl. Bact.* 24: 60–70.

Gibson T. and Stirling A.C. (1959). *The Bacteriology of Silage*. NAAS Quarterly Review, No. 44: 167–172.

Gilliland S.E. and Speck M.L. (1977). Deconjugation of bile acids by intestinal *Lactobacilli*. *Appl. Environ. Microbial.* 33: 15–18.

Gilliland S.E., Staley T.E. and Bush L.J. (1984). Importance of bile tolerance of *Lactobacillus acidophilus* used as a dietary adjunct. *J. Dairy Sci.* 67: 3045–3051.

Gilliland S.E., Nelson C.R. and Maxwell C. (1985). Assimilation of cholesterol by *Lactobacillus acidophilus*. *Appl. Environ. Microbiol.* 48:377–381.

Gilliland S.E. (1979). Beneficial inter-relationships between certain microorganisms and humans: candidate micro-organisms for use as dietary adjuncts. *J. Food Prot.* 42:164

Gilliland S.E. (1987). The importance of bile tolerance in *Lactobacilli* used as dietary adjuncts *Biotechnology in the Feed Industry*. (Ed T.P. Lyons) Alltech Technical Publications, Lexington, Kentucky.

Gilliland S.E. and Speck M.L. (1968). D-leucine as an auto-inhibitor of lactic *Streptococci*. *J. Dairy Sci.* 51:1573.

Gilliland S.E. and Speck M.L. (1977). Antagonistic action of *Lactobacillus acidophilus* towards intestinal and food-borne pathogens in associative culture. *J. Food Prot.* 40:829–823.

Girardeau J.P., Dubourguier H.C. and Contrepois M.G. (1980). Attachment des *Escherichia coli* enterophatho genes a la muqueuse intestinale. Bull. Groupe. *Tech. Vet.* 80: 4–B-190: 48–60.

Goepfert J.M. and Hicks R. (1969). Effect of volatile fatty acids on *Salmonella typhimurium*. *J. Bacteriol.* 97:956–962.

Goldberg H.S. (1964). Non-medical use of antibiotics. *Adv. in App. Micr.* 6, 91–117.

Goldin B.R. and Gorbach S.L. (1977). Alteration in fecal micro-flora enzymes related to diet, age, *Lactobacilli* supplements and dimethylhudrazine. *Cancer* 40: 2421–24 .

Goldin B.R. and Gorbach S.L. (1980). Effect of *Lactobacillus acidophilus* dietary supplements on 1,2 dimethylhydrazine dihydrochloride induced intestinal cancer in rats. *J. Natl. Cancer Inst.* 64:263.

Goldin B.R. and Gorbach S.L. (1977). Alterations in fecal micro-flora enzymes related to diet, age *Lactobacillus* supplements, and dimethylhydrazine. *Cancer* 40:2421.

Goldin B.R. and Gorbach S.L. (1984) The effect of milk and *Lactobacillus* feeding on human intestinal bacterial enzyme activity. *Am. J. Clin. Nutr.* 39: 756–761.

Goldstein I.J., Hughes R.C., Monsigny M., Osawa T. and Sharon N. (1980). What should be called a lectin? *Nature* (London). 285: 66.

Gombos S. (1991). The Hungarian experience of using biological products. The way forward to creating more market acceptable animal production. In. *Biotechnology in the Feed Industry* (Ed: T.P. Lyons). pp. 199–210. Alltech Technical Publications, Kentucky.

Goodenough E.R. and Kleyn D.H. (1976) Influence of viable yogurt micro-flora on digestion of lactose by the rat. *J. Dairy Sci.* 59: 701–705.

Goodling A.C., Cernglia G.J. and Herbert J.A. (1987). Production performance of white Leghorn layers fed *Lactobacillus* fermentation products. *Poultry Sci.* 66: 480TH.

Goodrich R.D., Garrett J.E., Gast D.R., Kirck M.A., Larson D.A. and Meiske J.C. (1984). Influence of monensin on the performance of cattle. *J.Anim. Sci.* 58:1484.

Gordon D., Macrae J. and Wheater D.M. (1957). A *Lactobacillus* preparation for use with antibiotics. *Lancet*, May 4: 899–901.

Gordon H.A. (1952). A morphological and biochemical approach. In: *Studies on the Growth Effect of Antibiotics in Germ Free Animals*. Lobund Inst., Univ. Notre Dame, South Bend, IN.

Gordon H.A. (1955). The role of the intestinal flora in absorption: a comparative study between germ-free and conventional animals. (Ed. T.Z Craky) *Intestinal Absorption and Malabsorption*. p 237. Raven Press, New York.

Gornall A.G., Bardawill C.J. and David M.M. (1949). Determination of serum proteins by means of the biuret reagent. *J. Biol. Chem.* 177: 751.

Gotz V., Romankiewicz J.A., Moss J. and Murrary H.W. (1979). Prophylaxis against ampicillin – associated diarrhoea with a *Lactobacillus* preparation. *Am. J. Hosp. Pharm.* 36:754–757

Graf W. (1983) Studies on the therapeutic properties of acidophilus milk. *Symp. Swedish Nutr. Found.* 15, 119–121.

Gribben B and Hughes G., (1989). Personal communication.

Grunewald K.K., Bruckner G.G., Mitchell G.E. Jr., Tucker R.E. and Schelling G.T. (1978). Gut microbial responses to *Lactobacilli*. *J. Anim. Sci.* 47 (Suppl. 1): 303.

Gualtieri M. and Betti S. (1985). Effect of *Streptococcus faecium* on sucking pigs. *Nutr. Abstr. Rev.* (Series B) 55: 34 .

Hale O.M. and Newton G.L. (1979). Effects of a non-viable *Lactobacillus* species fermentation product on performance of pigs. *J. Anim. Sci.* 48: 770–775.

Halliday L.J. (1985). Studies of rumen degradation in relation to the utilization of forage. Ph.D. Thesis, University of Wales.

Hamdan I.Y. and Mikolajcik E.M. (1973). Growth viability and antimicrobial activity of *Lactobacillus acidophilus*. *J. Dairy Sci.* 56:638. (Abstr).

Hamdan I.Y. and Mikolajcik E.M. (1974). Acidolin: an antibiotic produced by *Lactobacillus acidophilus*. *J. Antibiot.* 27:631–636.

Haresign and Ewing (1989). Review of probiotic products available in the UK. In: *Probiotics: Theory and Applications*. (Eds. B.A. Stark and J.M. Wilkinson). Chalcombe Publications, Wye College, England. 29–38.

Harker A.J. (1989). Probiotics and acidification as part of a 'natural' pig production programme. *The Feed Compounder.* 9:12–14.

Harper A.F., Kornegay E.T., Bryant K.L. and Thomas H.R. (1983). Efficiency of virginiamycin and a commercially available *Lactobacillus* probiotic in swine diets. *Anim. Feed. Sci. Technol.* 8: 69–76.

Harrison A.P. and Hansen P.A. (1950). A motile *Lactobacillus* from the caecal faeces of the turkey. *J. Bact.* 59:444–446.

Harrison G.A., Hemken R.W., Dawson R.A. and Harmon R.J. (1988). Influence of addition of yeast culture supplement to diets of lactacting cows on ruminal fermentation and microbial populations. *J. Anim. Sci.* 71:2967–2975.

Harvey J.D. (1992). Changing waste protein from a waste disposal problem to a valuable feed protein source: A role for enzymes in processing offal, feathers and dead birds. In: *Biotechnology in the Feed Industry* (Ed. T.P. Lyons). pp 109–119. Alltech Technical Publications, Nicholasville, Kentucky.

Hawley H.B., Shepard P.A. and Wheater D.M. (1959). Factors affecting the implantation of *Lactobacilli* in the intestine. *J. Appl. Bact.* 22: 360–367.

Hedde R.D. (1984). In: *Antimicrobials in Agriculture* (Ed M. Woodbine). pp. 359–368. Butterworths, London.

Hentges D.J., Que J.U., Casey S.W. and Stein A.J. (1984). The influence of streptomycin on colonization resistance in mice. *Microecol. Ther.* 14:53–63.

Hentges D.J. (Ed). (1983). *Human Intestinal Micro-flora in Health and Disease.* Academic Press, Inc., New York, NY.

Herrick J.B. (1972). Therapeutic nutrition using *Lactobacillus* species. *Vet. Med. and Small Anim. Clin.* 67:1249.

Hicks R.B., Gill D.R., Smith R.A. and Ball R.L. (1986). The effect of a microbial culture on health and performance of newly-arrived stocker cattle. *Okla. Agr. Exp. Sta. Res. Rep.* MP. 118: 256–259.

Hill I.R., Kenworthy R. and Porter P. (1970). Studies on the effect of dietary *Lactobacilli* on intestinal and urinary amines in pigs in relation to weaning and post-weaning diarrhoea. *Res. Vet. Sci.*, 11:320–326.

Hill I.R., Kenworthy R. and Porter P. (1970). The effect of dietary *Lactobacilli* on in-vitro catabolic activities of the small intestinal micro-flora of newly weaned pigs. *J. Med. Micr*, 3:593–605.

Hill M.J., Fadden K., Fernandez F. and Roberts A.K. (1986). Biochemical basis for microbial antagonism in the intestine. In: *Natural Antimicrobial Systems, Part I.* (Eds. G.W. Gould, M.E. Rhodes-Roberts, A.K. Charnley, R.M. Cooper and R.G. Board) pp. 29–39. Bath University Press, Bath, UK.

Hirsch A. and Grinsted E. (1951). The differentiation of the lactic *Streptococci* and their antibiotics. *J. Dairy Res.* 18:198–205.

Hirsh A. and Wheater D.M. (1951). The production of antibiotics by *Streptococci. J. Dairy Res.* 18:193–197.

Holden, P.J. (1976). The effect of commercial feed additives on the performance of early-weaned pigs. *Proc. Internat. Pig Vet. Soc:* AA5.

Hollister A., Cheeke P., Robinson K. and Palton N. (1991). Effect of dietary probiotics and acidifiers on performance of weaning rabbits. In: *Biotechnology in the Feed*

Industry. pp. 429–433. (Ed: T.P. Lyons). Alltech Technical Publications; Nicholasville, Kentucky.

Huddleson I.F., DuFrain J., Barrows K.C. and Giefel M. (1944). Antibacterial substances in plants. *Journal of the American Veterinary Medicine Association* 105: 394–412.

Hughes J. (1988). The effect of a high-strength yeast culture in the diet of early-weaned calves. *Anim. Prod.* 46: 526 (Abstr).

Hughes J. (1988). Calf, heifer and beef nutrition: designing tomorrow's feeds. In *Biotechnology in the Feed Industry*. pp. 67–78. (Ed: T.P. Lyons). Alltech Technical Publications, Nicholasville, Kentucky.

Humphrey T.J., Kirk J.A. and Cooper R.A. (1982). Effect of high acid milk replacer in conjunction with hay and concentrates on the faecal coliform population of pre-weaned calves. *Vet. Rec.* 110: 85.

Hungate R.E. (1966). *The Rumen and its Microbes*. Academic Press, New York.

Hurst A. (1981). Nisin. *Adv. Appl. Microbiol.* 27: 85–123.

Hutcheson D.P., Cole N.A., Keaton W., Graham G., Dunlap R. and Pittman K. (1980). The use of living, non-freeze-dried *Lactobacillus acidophilus* culture for receiving feedlot calves. *Proc Amer. Soc. Anim. Sci., West. Sect.* 31: 213–215.

Iglewski W.J. and Gerhardt N.B. (1978). Identification of an antibiotic-producing bacterium from the human intestinal tracts and characterisation of its antimicrobial product. *Antimicrobe Chemother.* 13 : 81–89.

Ingraham J.L., Maaloe O. and Neidhardt F.C. (1983). *Growth of the Bacterial Cell*. Sinauer Associates, Inc. Sunderland, Massachusetts.

Ingraham J.L. (1962) Temperature relationships. In: *The Bacteria*. (Eds: I.C Gunsalus and Stainier R.Y). Vol IV. pp. 265–296. Academic Press, New York and London.

Ingraham J.L. and Stokes J.L. (1959). Psychrophilic bacteria. *Bact. Rev.* 23, 97–108.

Inooka S., Uehara S. and Kimura M. (1986). The effect of *Bacillus natto* on the T and B lymphocytes from spleens of feeding chickens. *Poultry Sci.* 65: 1217–1219.

Inooka S. and Kimura M. (1983). The effect of *Bacillus natto* in feed on the sheep red blood cell antibody response in chickens. *Avian Dis.* 27: 1086–1089.

I.U.B. (1991). *Enzyme Nomenclature. Recommendations by the International Union of Biochemistry*. Academic Press, Orlando, Florida, U.S.A.

Jernigan M.A., Miles R.D. and Arafa A.S. (1985). Probiotics in poultry nutrition – a review. *World Poultry Sci.* J. 41: 99107.

Jervis H.R. and Biggers D.C. (1964) Mucosal enzymes in the cecum of conventional and germ free mice. *Anat. Rec*, 148:591–595.

Jezpwa L., Rafinski T.J. and Wrocinski T., (1966). Investigations on the antibiotic activity of *Allium sativum*. *Herba Polanica* 12:3–7.

Jones G.W. and Rutter J.M. (1972). Role of the K88 antigen in the pathogenesis of neonatal diarrhoea caused by *Escherichia coli* in piglets. *Inf. Imm.* 6:918,1972.

Jones R.J. and Megarrity R.G. (1986). Successful transfer of DHP degrading bacteria from Hawaiian goats to Australian ruminants to overcome the toxicity of *Leucaena*. *Australian Veterinary Journal* 63:259–262.

Jongbloed A.W. and Kemme P.A. (1990). Effect of pelleting mixed feeds on phytase activity and apparent absorbability of phosphorus and calcium in pigs. *Anim. Fd. Sci. Tech.* 28: 233–242.

Jonsson E. (1986). Persistence of *Lactobacillus* strain in the gut of sucking piglets and its influence on performance and health. *Swed. J. Agric. Res.* 16: 43–47.

Jonsson E. Bjorck L. and Claeson C.O. (1985). Survival of orally administered *Lactobacillus* strains in the gut of cannulated pigs. *Livest. Prod. Sci.* 12: 279–285.

Jonsson E. and Olsson L. (1985). The effect on performance, health and faecal microflora of feeding *Lactobacillus* strains to neonatal calves. *Swed. J. Agric. Res.* 15: 71–76.

Just A. (1983). The role of the large intestine in the digestion of nutrients and amino acid utilization in monogastrics. *IVth Int. Symp. Protein Metabolism and Nutrition.* pp. 289–309. Clermont-Ferrand, France.

Kasai K. and Kobayashi R. (1919). The stomach spirochete occurring in mammals. *J. Parasitol 6:1–11.*

Kay R.M. and Poole P. (1988a). *Enterococcus faecium* fed as a probiotic in the rearing of purchased calves. *Anim. Prod.* 46: 499 (Abstr).

Kay R.M. and Poole P. (1988b). *Lactobacillus acidophilus* fed as a probiotic in the rearing of purchased calves. *Anim. Prod.* 46: 525 (Abstr.).

Kenworthy R. and Crabb W.E. (1963). The intestine flora of young pigs with reference to early weaning and *Escherichia coli* scours. *J. Comp. Pathol,* 73: 215–228.

Ketaren P.P., Batterham E.S., Dettmann E.B. and Farrell D.J. (1993). Phosphorus studies in pigs. 3. Effect of phytase supplementation on the digestibility and availability of phosphorus in soya-bean meal for grower pigs. *British Journal of Nutrition.* 70:289–311.

Kiesling H.E. and Lofgreen G.P. (1981). Selected fermentation products for receiving cattle. *J. Anim. Sci.* 53 (Suppl. 1): 483.

Kiesling H.E. and Lofgreen G.P. (1982). A viable *Lactobacillus* culture for feedlot cattle. *J. Anim. Sci.* 55 (Suppl. 1):490.

Kim H.S. and Gilliland S.E. (1983). *Lactobacillus acidophilus* as a dietary adjunct for milk to aid lactose digestion in humans. *J. Dairy Sci.* 66:959.

Kingman S. (1993). Back to a plague-ridden future. *The Independent on Sunday,* United Kingdom. 28 February. 50

Kinsey C.M. (1980). Use of microbial additives in feed. A literature review. *Proc. 40th seminar meeting.* AFMA, *Nutr. Council,* San Antonio, Texas:25–30.

Kluber E.P., Pollmann D.S. and Blecha F. (1985). Effect of feeding *Streptococcus faecium* to artificially-reared pigs on growth, hematology and cell-mediated immunity. *Nutr. Rpt. Intern.* 32: 57–56.

Knox K.W. and Wickes A.J. (1973). Immunological properties of teichoic acids. *Bact. Rev.* 37:215–257.

Kodama R. (1952) Studies on lactic acid bacteria II. Lactolin, a new antibiotic substance produced by lactic acid bacteria. *J. Bact.* 5:72–77.

Kopeloff N. (1926). *Lactobacilus acidophilus,* Williams & Williams, Baltimore.

Kopeloff N. and Beerman P.L. (1925). *Lactobacillus acidophilus* vs *L bulgaricus* milk feeding. *Proc. Soc. Exp. Biol. and Med,* 22:318.

Korzybski T., Kowszyk-Gindifer Z. and Kurylowiez W. (1978). *Antibiotics: Origin, Nature and Properties, Vol. III*. Pergamon Press, Oxford.

Kotarski S.F. and Savage D.C. (1979). Models for study of the specificity by which endogenous *Lactobacilli* adhere to murine gastric epithelia. *Infect. Immun.* 26:966.

Kroulik J.T., Birkey L.A. and Wiseman H.G. (1955). The microbial populations of the green plant and of the cut forage prior to ensiling. *J. Dairy Sci.* 38: 256–262.

Lacey R.W. (1983) In *Antimicrobials in Agriculture* (Ed: M Woodbine). Butterworths, London. 17: 221–235.

Lacey R.W. (1988). Rarity of tylosin resistance in pathogenic bacteria. *Vet Rec.* 122. 438–439.

Laforge R.R. and Pollmam D.S. (1984). Effect of *Bacillus subtilis* on sow and baby pig performance and fecal bacteria populations. *J. Anim. Sci.* 59 (Suppl. 1): 248.

Langston C.W., Bouma C. and Conner R.M. (1962). Chemical and bacteriological changes in grass silage during the early period of fermentation. Bacteriological changes. *J. Dairy Sci.* 45: 618–624.

Larue A. (1960). Effects of acidified milk on intestinal pathogens. *Can. Med. Ass. J.* 83: 1002–1004.

Lassiter C.A., Grimes R.M. and Duncan C.W. (1958), Influence of antibiotics on growth and protein metabolism of young dairy calves. *J.Dairy Sci*, 41:1416.

Lassiter C.A. (1955). Antibiotics as growth stimulants for dairy cattle. A review. *J. Dairy Sci., 38:1102*.

Ledinek M. (1970). Studies on participation of micro-organisms in digestive processes of pigs fed raw and steamed potatoes. Thesis, Ludwig-Maximilians-Universitat, Munchen.

Lee A. and Hazell S.L. (1988). *Campylobacter pylori* in health and disease: an ecological perspective. *Microb. Ecol. Health Dis.* 1:1–16.

Lee A., Hazell S.L., O'Rourke J. and Kouprach S. (1988). Isolation of a spiral-shaped bacterium from the cat stomach. *Infect. Immun.* 56:2843–2850.

Lessard M. and Brisson G.J. (1987). Effect of a *Lactobacillus* fermentation product on growth, immune response and faecal enzyme activity in weaned pigs. *Can. J. Anim. Sci.* 67: 509–516.

Lev M. and Forbes M. (1959). Growth response to dietary penicillin of germ-free chicks and of chicks with a defined intestinal flora. *Brit. J. Nutr.* 13: 78–84.

Levensen S.M., Crawley L.V., Horowitz R.E. and Malm O.J. (1959). The metabolism of carbon-labelled urea in the germ-free rat. *J. Biol. Chem*, 234:2061–2062.

Levin R.J. (1979) Fundamental concepts of the structure and function of the gastrointestinal epithelium. In: *Scientific Basis of Gastroenterology* (Ed: H.L. Duthie), Churchill Livingstone, Edinburgh. pp. 307–351.

Lilley D.M. and Stillwell R.H. (1965). Probiotics: Growth-promoting factors produced by micro-organisms. *J. Bact.* 89:747–748.

Lincoln R.E. (1960). Control of stock culture preservation and inoculum build-up in bacterial fermentation. *J.Biochem. Microbiol. Tech. Eng.* 2:481–500.

Lindsey T.O., Hedde R.D. and Sokolek J.A. (1985). Characterisation of feed additive effects on the gut micro-flora of chickens. *Poultry Sci.* 64(Suppl 1):27–28 (Abstr).

Linton A.H. and Hinton M.H. (1988). *Enterobacteriacae* associated with animals in health and disease. *J. Appl. Bact.* Symposium Supplement. 715–855.

Lloyd A.B., Cummings R.B. and Kent R.D. (1977). Prevention of *Salmonella typhimurium* infection in poultry by pretreatment of chicks and poults with intestinal extracts. *Austral. Vet. J.* 53:82–87.

Loveland J., Kesler E.M. and Doores A. (1983). Fermentation of a mixture of waste milk and colostrum for feeding young calves. *J. Dairy Sci.* 66: 1312–1318.

Lubis, D. (1983). The antibacterial activity of yoghurt cultures towards *Salmonella typhimirium. Dairy Sci. Abst.* 45: nos. 5044 and 5047.

Luckey T.D. (1963). *Germfree Life and Gnotobiology*. Academic Press, New York.

Lyons T.P. (1986). Biotechnology in the feed industry. In: *Second Annual Biotechnology Symposium*. (Ed: T.P. Lyons) Alltech Inc. Nicholasville, Kentucky.

Machado-Neto R, Graves C.N. and Curtis S.E. (1987). Immunoglobins in piglets from sows heat-stressed prepartum. *J. Anim. Sci* 65:445–455.

Macleod R.A. and Snell E.E. (1947). Some mineral requirements of the lactic acid bacteria. *J. Biol. Chem.* 170:351–353.

MAFF (1994). Identification Notes Issued by Chemical Safety Food Division, MAFF, Room 531, Ergon House, c/o Nobel House, (7 Smith Square, London. SW1P 3JR).

Mahan D.C. and Newland H.W. (1976). Short term effects of the addition of oats, bacterial cultures (probiotic), and antibiotics to the diets of weaned pigs. *Ohio Agr. Res. Bulletin*: 35.

Mann G.V. (1977) A factor in yogurt which lowers cholesteremia in man. *Atherosclerosis* 26:335–340.

Manten A. and Meyerman-Wisse J. (1962). A systematic study of antibiotic antagonism. Antonie van Leeuwenhoek. *J. Microbiol. Serol.* 28:321–345.

March B.E., Soong R. and Macmillan C. (1978). Growth rate, feed conversion and dietary metabolisable energy in response to virginiamycin supplementation of different diets. *Poultry Sci.* 57:1346–1350.

March B.E. (1979). The host and its micro-flora - an ecological unit. *J. Anim. Sci.*, 49:857–867.

Marshall V.M., Philips S.M. and Turvey A. (1982). Isolation of a hydrogen peroxide – producing strain of *Lactobacillus* from calf gut. *Res. Vet. Sci.* 32: 259–260.

Mason V.C. (1980). Role of the large intestine in the processes of digestion and absorption in the pig. In: *Current Concepts of Digestion and Absorption in Pigs*. pp. 112–129. (Eds. A.G Low and I.G. Partridge), NIRD, Reading.

Mattick A.T.R. and Hirsch A. (1944). A powerful inhibitory substance produced by group N *Streptococci. Nature*, 154:551.

Maxwell C.V., Buchanan D.S., Owens F.N., Gilliland S.E.and Luce W.G. (1982). Effect of probiotic supplementation on performance, faecal parameters and digestibility in growing – finishing pigs. *J. Anim. Sci.* 55 (Suppl. 1): 284.

Mayberry W.R., Prochazka G.J. and Payne W.J. (1967). Growth yields of bacteria on selected organic compounds. *Appl. Microbiol.* 15:1332–1338.

Mayer, J.B. (1962). Possibilite d'une Therapie Physiologiquement Antibiotique Chez La Nourisson. Xth International Congress of Paediatrics, Lisbon.

McAllister J.S., Kurtz H.J. and Short E.C. (1979). Changes in the intestinal flora of young pigs with post-weaning diarrhoea or odema disease. *J. Anim. Sci.* 49: 868–879.

McCormick M.E. (1984). Probiotics in ruminant nutrition and health. In: *Proc. 1984 Georgia Nutrition Conference for the Feed Industry.*

McCormick E.L. and Savage D.C. (1983). Characterization of *Lactobacillus sp.* strain 100–37 from the murine gastrointestinal tract: ecology, plasmid content and antagonistic activity toward *Clostridium ramosum. Appl. Environ. Microbiol.* 46:1103–1112.

McDonald P., Edwards R.A. and Greenhalgh J.F.D. (1988). *Animal Nutrition (4th edition).* Longman Group Scientific and Technical; Harlow, England.

Mead G.C. and Adams B.W. (1975). Some observations on the caecal micro-flora of the chick during the first two weeks of life. *Br. Poult. Sci.*, 16: 169–176.

Mead G.C. and Impey C.S. (1984). Control of *Salmonella* colonization in poultry flocks by defined gut-flora treatment. (Ed. GH Snoeyenbos). In: *Proceedings of the International Symposium on Salmonella* pp. 72–79. American Association of Avian Pathologists, Kennett Square, PA.

Mead G.C. and Impey C.S. (1987) In *Elimination of Pathogenic Organisms from Meat and Poultry.* (Ed F.J.M. Smulders), pp. 57–77. Elsevier Science Publishers, Amsterdam.

Mepham T.B. (1993). Biotechnology – the ethics. *Anim. Prod.* 57:353–359.

Metchnikoff E. (1907). *Prolongation of Life.* G.P. Putnam and Sons, New York.

Metchnikoff E. (1908). *The Prolongation of Life.* London: Heinemann.

Meynell G.G. (1963). Antibacterial mechanisms of the mouse gut. II. The role of volatile fatty acids in the normal gut. *Br. J. Exp. Path.* 44:209–213.

Meyrath J. and Suchanek G. (1972) Inoculation techniques – effects due to quality and quantity of inoculum. In *Methods in Microbiology* Vol. 7B. pp, 159–209. (Eds J.R. Norris and D.W. Ribbons) Academic Press, London.

Michel M. (1968). Degradation of arginine by the microbial flora of the pig. I. Study in vitro. *Ann. Biol. Anim. Biochem. Biophys.* 8: 385.

Mikolajcik E. M. and Hamdan I.Y. (1975). *Lactobacillus acidophilus* 11: Antimicrobial agents. *Cult. Dairy Prod. J.* 10: 18.

Miles A.A. and Misra S.S. (1938). The estimation of the bacteriocidal power of the blood. *J. Hyg. Camb.* 38: 732.

Miles R.D., Arafa A.S., Harms R.H., Carlson C.W., Reid B.L. and Crawford J.S. (1981a). Effects of living non-freeze-dried *Lactobacillus acidophilus* culture on performance, egg quality and gut micro-flora in commercial layers. *Poultry Sci.* 60: 993–1004.

Miles R.D., Wilson H.R., Arafa A.S., Coligado E.C. and Ingram D.R. (1981b). The performance of Bobwhite Quail fed diets containing *Lactobacilli. Poultry Sci.* 60: 894–896.

Miles R.D., Wilson H.R. and Ingram D.R. (1981c). Productive performance of Bobwhite Quail breeders fed a diet containing a *Lactobacillus* culture. *Poultry Sci.* 60: 1581–1582.

Miles R.D. (1993). Manipulation of micro-flora in the digestive tract : natural ways to prevent colonisation by pathogens. In. *Biotechnology in the Feed Industry* (Ed T.P. Lyons) Alltech Technical Publications, Nicholasville, Kentucky.

Miller P.C. and Bohnhoff M. (1963). Changes in the mouse's enteric micro-flora associated with enhanced susceptibility of Salmonella infection following streptomycin treatment. *J. Inf. Dis.* 112:59–62.

Mitchell I.D.G. and Kenworthy R. (1976). Investigations on a metabolite from *Lactobacillus bulgaricus* which neutralizes the effect of enterotoxin from *Escherichia coli* pathogenic for pig. *J. App. Bact*, 41:163–174.

Mongin P. (1981). Recent advances in dietary anion-cation balance in poultry. In: *Recent Advances in Animal Nutrition* (1981) (Ed W. Haresign). pp. 109–119. Butterworths, London.

Moon H.W. (1975). An evaluation of the evidence of preventative or therapeutic effects of *Lactobacilli* in enteric *E.coli* infection of swine. *Amer. Feed Manufacturers Assoc. Nutr Council*, November. p. 42.

Moore E. (1993) Analytical and regulatory requirements for microbia products. The use of DNA fingerprinting and biotechnology to ensure presence and survival through feed processing. In *Biotechnology in the Feed Industry* (Ed T.P. Lyons) pp. 245–254. Alltech Technical Publications, Nicholasville, Kentucky, US.

Moore P.R., Evenson A., Kuckey T.D., McCoy E., Elvehjem C.A. and Hart E.B. (1946). Use of sulfasuxidine streptothricin and streptomycin in nutritional studies with the chick. *J. Bio.Chem.* 165:437–441.

Moore W.E.C. and Holdeman L.V. (1974). Human faecal flora: The normal flora of 20 Japanese-Hawaiians. *Appl. Microbial.* 27: 961–979.

Morishita Y., Mitsuoka T., Kaneuchi I.C., Yamamota and Ogata M. (1971). Specific establishment of *Lactobacilli* in the digestive tract of germ-free chickens. *Jpn J. Microbiol.* 15: 531–538.

Morland B. and Midtvedt T. (1984). Phagocytosis, peritoneal influx and enzyme activities in peritoneal macrophages from germfree, conventional and ex-germfree mice. *Infect. Immun.* 44:750–752.

Morrill J.L., Dayton A.D. and Mickelsen R. (1977). Cultured milk and antibiotics for young calves. *J. Dairy Sci.* 60: 1105–1109.

Morris J.A., Sojka W.J. and Ready R.A. (1985). Serological comparison of the *Escherichia coli* prototype strains for the F(Y) and Att25 adhesins implicated in neonatal diarrhoea in calves. *Res. Vet. Sci.* 38: 246–247.

Mould F.L. and Orskov E.R. (1983/84). Manipulation of rumen fluid pH and its influence on cellulolysis *in sacco*, dry matter degradation and the rumen micro-flora of sheep offered either hay or concentrate. *Anim. Feed Sci. Technol.* 10: 1–14.

Muller R. and Kieslich K. (1966) Technology of the microbiological preparation of organic substances. *Angwante Chem. (International Edition)* 5: 653–662.

Muralidhara K.S., Sheggeby G.G., Elliker P.R. and England D.C. (1972). Effects of feeding *Lactobacillus* concentrates on coliform excretion and scouring in swine. *J. Dairy Sci.* 55:635–640.

Muralidhara K.S., Sheggeby G.G., Elliker P.R., England D.C. and Sandine W.E. (1977). Effect of feeding *Lactobacilli* on the coliform and *Lactobacillus* flora of intestinal tissue and faeces from piglets. *J. Food Prot.* 40:288–295.

Muralidhara K.S., Sandine W.E., England D.C. and Elliker P.R. (1973). Colonization of *Escherichia coli* and *Lactobacillus* in intestines of pigs. J. *Dairy Sci.* 56:635–642.

Muralidhara K.S. (1974) Effect of feeding concentrates of *Lactobacillus* organisms on intestinal colonisation by *Escherichia coli* in swine. Thesis, Oregon State University.

Murumatsu T., Tohasu O., Furuse M., Takaski I. and Ilkumura J. (1987). Influence of the gut micro-flora on protein synthesis in tissues and in the whole body of chicks. *Biochem. Journ.* 246: 475–479.

Mushin R. and Dubos R. (1965). Colonization of the mouse intestine with *Escherichia coli. J. Exptl. Med.*, 122:745–752.

Nahaisi M.H. (1986) *Lactobacillus acidophilus,* therapeutic properties, products and enumaration. *Dev. Food Micr.* 2:153–178.

Nakagawa J., Tamaki S., Tomioko S. and Matsuhashi M. (1984) Functional biosynthesis of cell wall peptidoglycan by polymorphic bifunctional polypeptides *J. Biol. Chem,* 259:13937–13943.

Nelson T.S., Shieh T.R., Wodzinskir J., and Gware J.H. (1968). *Poult. Sci.* 47, pp. 1842–1848.

Neut E., Bezirtzoglou A., Tomond C., Beerens H., Delcroix M., and Noel A.M. (1987). Bacterial colonization of the large intestine in newborns delivered by caesarean section. *Zbl. Bakt., Hyg.* A, 266:330–337.

Niv M. (1963) Yoghurt in the treatment of infantile diarrhoea. *Clin. Paed,* 2: 407–411.

Ofek I., Mirelman D. and Sharon N. (1977). Adherence of *Escherichia coli* to human mucosal cells mediated by mannose receptors. *Nature* 256: 623–625.

Okumura J., Hewitt D., Salter D.N. and Coates M.E. (1976). The role of the gut micro-flora in the utilisation of dietary urea by the chick. *Brit. J. Nutr.* 36:265–272.

Op Den Camp H.J.M., Oosterhof A. and Veerkamp J.H. (1985). Interaction of bifidobacterial lipoteichoic acid with human intestinal epithelial cells. *Infect. Immun.* 47:332–334.

Orla-Jensen S. (1943). *The Lactic Acid Bacteria.* Copenhagen: Ejnar Munksgaard.

Oxford A.E. (1944). Diplococcin, an anti-bacterial protein elaborated by certain milk *streptococci. Biochem J,* 38:178–182.

Oyofo B., Deloach J.R., Corrier D.E., Norman J.O., Ziprin R.I. and Mollenhauer H.H. (1989). Prevention of *Salmonella typhimurium* colonisation of broilers with D-mannose. *Poult. Sci.* 68: 1357–1360.

Ozawa K., Yabu-uchi K., Yamanaka K., Yamashita Y., Nomura S. and Oku I. (1983). Effect of *Streptococcus Faecalis* BIO-4R on intestinal flora of weaning piglets and calves. *Appl. Environ. Microbial.* 45: 1513–1518.

Ozawa K., Yokota H., Kimura M. and Mitsuoka T. (1981). Effects of administration of *Bacillus subtilis* strain BN on intestinal flora of weaning piglets. *Jpn. J. Vet. Sci.* 43: 771–775.

Pagan J.D. (1990). Effect of yeast culture supplementation in mature horses. *J. Anim, Sci,* 68.

Page T.G., Ward T.L. and Southern L.L. (1991). Effects of chromium picolinate on growth on carcass characteristics of growing finishing pigs. *J. Anim. Sci.* 69. (Suppl 1):356.

Parker R.B. (1974). Probiotics, the other half of the antibiotic story. *Anim. Nutr. Health*, 29: 4–8.

Parra R. (1978). Comparison of foregut and hindgut fermentation in herbivores. *In: Ecology of Arboreal Folivores*. (Ed: G.G. Montgomery), Smithsonian Institution Press, Washington D.C.

Patience J.F. (1989). The physiological basis of electrolytes in animal nutrition. In: *Recent Advances in Animal Nutrition* – 1989. (Eds: W. Haresign and D.J.A. Cole) pp. 212. Butterworths, London.

Patrick T.E., Collins J.A. and Goodwin T.L. (1973). Isolation of *Salmonella* from carcasses of steam and water scalded poultry. *J. Milk & Food Tech.* 36: 34–36.

Paul D. and Hoskins L.C. (1972) Effect of oral *Lactobacillus* feeding on fecal *Lactobacillus* counts. *Am. J. Clin. Nutr.* 25: 763–765.

Perdigon G., Nader de Macias M.E., Alvarez S., Oliver G., and Pesce de Ruiz Holgado A.A. (1986a). Effect of orally administered *Lactobacilli* on macrophage activation in mice. *Infect. Immun.* 53:404–410.

Perdigon G., Nader de Macias M.E., Alvarez S., Medici M., Oliver G. and Pesce de Ruiz Holgado A.A. (1986b). Effect of a mixture of *Lactobacillus casei* and *Lactobacillus acidophilus* administered orally on the immune system in mice. *J. Fd. Protect*, 49:986–989.

Perdigon G., Alvarez S., Nader de Macias M.E., Margni R.A., Oliver G. and Pesce de Ruiz Holgado A.A. (1986). *Lactobacilli* administered orally induce release of enzymes from peritoneal macrophages in mice. *Milchwis.* 41:344–348.

Pettersson D., Graham H. and Aman P. (1990). Enzyme supplementation of broiler chicken diets based on cereals with endosperm cell walls rich in arabinoxylans or mixed-linked β–glucans. *Anim. Prod.* 51:201–207.

Pivnick H. and Nurmi E. (1982) In *Developments in Food Microbiology* –1: (Ed: R. Davies). Applied Science Publishers, London. pp. 41–70.

Pohl P., Lintermans P., Van Muylem K. and Schotte M. (1982). *Colibacillies enterotoxigenes* de veau possedant un antigene d'attachement différent de l'antigene K99. *Ann. Med. Vet.* 126: 569–571.

Pohl P., Lintermans P. and Van Muylem K. (1984). Freguence des adhesines K99 et Att25 chez les *E.coli* du veau. *Ann. Med. Vet.* 128: 555–558.

Pollman D.S., Danielson D.M. and Peo E.R. (1980). Effects of microbial feed additives on performance of starter and growing-finishing pigs. *J. Anim Sci.*, 51:577.

Pollman D.S. (1986). Non nutritive feed additives – what are they. *Proceedings of the Twenty Second Annual Nutrition Conference for Feed Manufacturers*, Toronto, Canadian Feed Industry Association. pp. 122–136.

Pollmann D.S., Kennedy G.A., Koch B.A. and Allee G.L. (1984a) Influence of non-viable *Lactobacillus* fermentation product on artificially reared pigs. *Nutr. Rep. Intern.* 29: 977–982.

Pollmann D.S., Johnston M.E., Allee G.L. and Hines R.H. (1984b). Effect of *Bacillus subtilis* addition to Carbadox – medicated starter pig diets. *J. Anim. Sci.* 59 (Suppl. 1):275.

Pollmann D.S. (1986). Probiotics in pig diets. In *Recent Advances in Animal Nutrition* (Eds: Haresign W. and Cole D.J.A.) : 193–205. London Butterworths.

Pollmann D.S. and Bandyk, C.A. (1984). Stability of viable *Lactobacillus* products. *Anim. Feed Sci. Tech.* 11: 261–267.

Pollmann D.S., Danielson D.M. and Peo Jr. E.R. (1980a). Effects of microbial feed additives on performance of starter and growing – finishing pigs. *J. Anim. Sci.* 51: 577–581.

Pollmann D.S., Danielson D.M. and Peo Jr. E.R. (1980b). Effect of *Lactobacillus acidophilus* on starter pigs fed a diet supplemented with lactose. *J. Anim. Sci.* 51: 638–644.

Pollmann D.S., Danielson D.M., Wren W.B., Peo Jr. E.R. and Shahani K.M. (1980c). Influence of *Lactobacillus acidophilus* inoculum on gnotobiotic and conventional pigs. *J. Anim. Sci.* 51: 629–637.

Porter W.L. (1989) Practical applications of probiotics in livestock production. In: *Probiotics: Theory and Applications*. (Eds: B.A. Stark and J.M. Wilkinson). Chalcombe Publications, Wye College, England. pp. 39–45.

Porter P., Powell J.R., Allen W.D. and Linggood M.A. (1985). In: *Virulence of Escherischia coli*. pp. 271–287. (Ed: M. Sussman). Academic Press, London.

Porter P. and Kenworthy R. (1969). A study of intestinal and urinary amines in pigs in relation to weaning. *Res. Vet. Sci.* 10:440.

Porter W.L. and Ewing W.N. (1988). Unpublished data.

Porter W.L. (1986). Personal communication.

Powles J. and Cole D.J.A. (1993). Research examines use of lactose in young pigs' diets. *Feedstuffs*, 65(8) 13–16.

Price R.J. and Lee J.S. (1970) Inhibition of *Pseudomonas* species by hydrogen peroxide producing *Lactobacilli J. Milk Food Technol*, 33: 13–18.

Pruit K.M. and Reiter B. (1985). *The Lactoperoxidase System* (Marce Dekker). 147–149.

Pugh R. (1992). Evaluation of Allzyme BG (Beta glucanase) supplementation of broiler diets under commercial production conditions. In: *Biotechnology in the Feed Industry:* (Ed: T.P. Lyons) Alltech Technical Publications, Nicholasville. Kentucky. pp. 60.

Pusztai A., Grant G., King T.P. and Clarke E.M.W (1990). Chemical Probiosis. In *Recent Advances in Animal Nutrition* (Eds: W. Haresign and D.J.A. Cole). Butterworths, London. pp. 47–60.

Pusztai A. (1986). The biological effects of lectins in the diet of animals and man. In: *Lectins: Biology, Biochemistry, Clinical Biochemistry* (Eds: T.C. Bog-Hansen and E. Van Driessche) 5: 317–327.

Radisson J.J., Smith C.K. and Ward G.M. (1956). The mode of action of antibiotics in the nutrition of the dairy calf. 1. Effect of terramycin administered orally on the performance and intestinal flora of young dairy calves. *J. Dairy Sci.* 39:1260–1264.

Rantala M. and Nurmi E. (1973). Prevention of the growth of *Salmonella infantis* in chicks by the flora of the alimentary tract of chickens. *Br. Poult. Sci.* 14: 627–630.

Ratcliffe B. (1985) In: *Digestive Physiology in the Pig* (Eds: A. Just, H. Jorgensen and J.A. Fernandez). Copenhagen, National Institute of Animal Science, Report No. 580 pp. 245–267.

Reddy G.V., Friend B.A., Shahani K.M. and Farmer R. (1983). Antitumor activity of yoghurt components. *J. Food Protect*, 46: 8–11.

Redmond H.E. and Moore R.W. (1965). Biological effect of introducing *Lactobacillus acidophilus* into swine herd experiencing enteritis. *The Southwestern Veterinarian* 287–288.

Rees L.P., Minney S.F., Pummer N.T., Slater J.H, and Skyrme D.A. (1993). A qualitative assessment of the anti-microbial activity of garlic *(Allium sativum)*. *World Journal of Microbiology and Biotechnology*. 9:303–307.

Rettger L. (1929). Some aspects of intestinal bacteriology in relation to health. *Am. J. Publ. Health*. 19:771–776.

Rettger L.F., Levy M.N., Weinstein L. and Weiss J.E. (1935). Lactobacillus acidophilus *and its therapeutic application*. Yale University Press, New Haven, Connecticut.

Rettger L.F. and Cheplin H.A. (1921a). Therapeutic application of *Bacillus acidophilus Proc. Soc. Exp. Biol. Med*, 19:72–75.

Roach S., Savage D.C. and Tannock G.W. (1977). *Lactobacilli* isolated from the stomach of conventional mice. *App. Env. Microbiol*, 33:1197.

Robbins S. (1987). Probiotics. A European Perspective. *The Feed Compounder* HGM Publications, Derby, England. 7: 24–28.

Robinson I.M., Whipp S.C., Bucklin J.A. and Allison M.J. (1984). Characterization of predominant bacteria from the colons of normal and dysenteric pigs. *Appl. Envr. Micr*, 48:964–969.

Robinson J.J. and McEvoy T.C. (1993). Biotechnology – the possibilities. *Anim. Prod.* 1993. 57:335–352.

Rogers L.A. (1928). The inhibiting effect of *Streptococcus lactis* on *Lactobacillus bulgaricus*. *J. Bacteriol.* 16: 321–325.

Rogers G.E. (1990) Improvement of wool production through genetic engineering. *Trends in Biotechnology*. 8:6–11.

Rogosa M., Franklin J.G. and Perry K.D. (1961). Correlation of the vitamin requirements with cultural and biochemical characters of *Lactobacillus sp. J. Gen. Micr.* 25. 473–482.

Rogosa M., Mitchell J.A. and Wiseman R.F. (1951a). A selective medium for the isolation and enumeration of oral and faecal *Lactobacilli*. *J. Bact.* 62: 132–134.

Rogosa M., Mitchell J.A. and Wiseman R.F. (1951b). A selective medium for the isolation and enumeration of oral *Lactobacilli*. *J. Dent. Res*, 40:682–683.

Rosebury T. (1962). *Micro-organisms Indigenous to Man*. McGraw-Hill Book Co. Inc. New York. p4 35.

Rosell V. (1987). Acidification and probiotics in Spanish pig and calf rearing. In: *Biotechnology in the Feed Industry*. pp. 177–180. (Ed: T.P. Lyons). Alltech Technical Publications. Nicholasville, Kentucky.

Roth R.X. and Kirchgessner M. (1987). Nutritive effects of *Streprococcus faecium* M-74 in starter pigs. *Nutr. Abst. Rev. (Series B)* 57: 364.

Rozee K.R., Cooper D., Lam D. and Costerton J.W. (1982). Microbial flora of the mouse ileum mucous layer and epithelial surface. *Appl. Env. Micr*, 43:1451–1463.

Rubin H.E. and Vaughan F. (1979). Elucidation of the inhibitory factors of yoghurt against *Salmonella typhimurium*. *J. Dairy Sci.* 62: 1873.

Ruppin H., Bar-Meir S., Soergel K.H., Wood C.M. and Schmitt M.G. (1980). Absorption of short-chain fatty acids by the colon. *Gastroenterol*, 78:1500–1507.

Rusoff L.L., Cummings A.H., Stone E.J. and Johnston J.E. (1959). Effect of high-level administation of chlortetracycline at birth on the health and growth of young dairy calves. *J.Dairy Sci.* 42:856–858.

Russel E.G. (1979). Types and distribution of anaerobic bacteria in the large intestine of pigs. *Appl. Env. Micr,* 37:187–193.

Russell J.B., Sharp W.M. and Baldwin R.L. (1989). The effect of pH on maximum bacterial growth rate and its possible role as a determinant of bacterial competition in the rumen. *J. Anim Sci.* 48: 251–255.

Saito H., Tomioka H. and Nagashima K. (1987). Protective and therapautic efficacy of *Lactobacillus casei* against experimental murine infections due to *Mycobacterium fortuitum* complex. *J. Gen. Microbiol.* 133: 2843–2851.

Salanitro J.P., Blake I.G. and Muirhead P.A. (1974a). Studies on the cecal micro-flora of commercial broiler chickens. *Appl. Micr.* 28: 439–447.

Salter D.N. (1973). The influence of gut micro-organisms utilisation of dietary protein. *Proc. Nutr. Soc.* 32: 65–71.

Sandine W.E., Muraldihara K.S., Elliker P.R. and England D.C. (1972). Lactic acid bacteria in food and health: a review with special reference to enteropathogenic *Escherichia coli* as well as certain enteric diseases and their treatment with antibiotics and *Lactobacilli. J. Milk Fd. Tech.,* 35: 691–702.

Sandine W.E. (1979). Roles of *Lactobacillus* in the intestinal tract. *J. Food Protect.* 42: 259–262.

Savage D.C. (1979). Introduction to mechanisms of association of indigenous microbes. *Am. J. Clin, Nutr.* 32: 113–118.

Savage D.C., Dubos R.J. and Schaedler R.W. (1968) The gastrointestinal epithelium and its autochthonous bacterial flora. *J. Exp. Med, 127:67–76.*

Savage D.C., Siegel J.E., Snellen J.E. and Whitt D.D. (1981). Transit time of epithelial cells in the small intestines of germ-free mice and ex-germfree mice associated with indigenous micro-organisms. *Appl. Env. Micr,* 42:996–1001.

Savage D.C. (1969). Microbial interference between indigenous yeast and *Lactobacilli* in the rodent stomach. *J. Bact,* 98: 1278–1283.

Savage D.C. (1972). Associations and physiological interactions of indigenous micro-organisms and gastro-intestinal epithelia. *Amer. J. Clin. Nutr.* 25: 1372–1379

Savage D.C. (1977) Microbial ecology of the gastro-intestinal tract. *Ann. Rev. Micr,* 31: 107–133.

Savage D.C. (1980). Adherence of normal flora to mucosal surfaces. – In *Bacterial Adherence* (Ed: E.H. Beachey): 33–59. Chapman and Hall; London.

Savage D.C. (1981). Mode of action and potential of probiotics. *Proc. Flor. Nutr. Conf.:* 3–38, U.S.A.

Savage D.C. (1983). Mechanisms by which indigenous microorganisms colonize gastro-intestinal epithelial surfaces. *Prog. Fd. Nutr. Sci,* 7:65–74.

Savage D.C. (1984). Activities of micro-organisms attached to living surfaces. In *Microbial Adhesion and Aggregation.* (Ed: K.C. Marshall), Springer-Verlag, Berlin. 233–249.

Savage D.C. (1984). Adherence of the normal flora. In *Attachment of Organisms to the Gut Mucosa.* Vol. 1. (Ed: E.C. Boedeker) 3–10. CRC Press, Florida.

Savage D.C. (1985). Effects of host animals of bacteria adhering to epithelial surfaces. In *Bacterial Adhesion*, (Eds: D.C. Savage and M. Fletcher), Plenum, New York. 437–463

Savage D.C. (1986). Gastro-intestinal micro-flora in mammalian nutrition. *Ann. Rev. Nutr.* 6:155–178.

Savage D.C. (1987). Factors influencing bio-control of bacterial pathogens in the intestine. *Fd. Technol.* 7:82–87.

Savage D.C. (1989). The ecological digestive system and its colonisation. *Rev. Sci. Tech. Off. Int. Epiz.* 8:259–273.

Savage D.C. and Dubos R.J. (1967), Localisation of indigenous yeast in the murine stomach. *J. Bact*, 94: 1811–1816.

Savage D.C. and Whitt D.D. (1982). Influence of the indigenous microbiota on amounts of protein, DNA, and alkaline phosphatase activity extractable from epithelial cells of the small intestines of mice. *Inf. Immun*, 37:539–549.

Schaedler R.W., Dubos R. and Costello R. (1965). The development of the bacterial flora in the gastro-intestinal tract of mice. *J. Exp. Med*, 122: 59–66.

Schaedler R.W. and Dubos R.J. (1962). The fecal flora of various strains of mice. Its bearing on their susceptibility to endotoxin. *J. Exp. Med.* 115: 1149–1160.

Schindler J. and Schmid R.D. (1982). Fragrance of aroma chemicals-microbial synthesis and enzymatic transformation – a review. *Process. Biochem*, 17:2–8.

Schleifer K.H. and Kandler O. (1972) Peptidoglycan types of bacterial cell walls and their taxonomic implications. *Bacteriol Rev.* 36: 407–477.

Schulman A. (1973). Effect of weaning on pH changes of the contents of the piglet's stomach and duodenum. *Nord. Vet. Med.* 25:220.

Schwab C.G., Moore J.J., Hoyt P.M. and Prentice J.L. (1980). Performance and faecal flora of calves fed a nonviable *Lactobacillus bulgaricus* fermentation product. *J. Dairy Sci.* 63: 1412–1423.

Schwarz K. and Merz W. (1957). A glucose tolerance factor and its differentiation from factor 3. *Arch. Biochem. Biophys.* 72: 515.

Setoyama T., Nomoto K., YokokuraT. and Mutai M. (1985). Protective effect of lipoteichoic acid from *Lactobacillus casei* and *Lactobacillus fermentum* against *Pseudomonas aeruginosa* in mice. *J. Gen. Micro.* 131:2501–2503.

Shahani K.M., Vakıl J.R. and Kilara A. (1976). Natural antibiotic activity of *Lactobacillus acidophilus* and *bulgaricus*. I. Cultured conditions required for the production of antibiosis. *Cult. Dairy Prod. J.* 11:14.

Shahani K.M. and Ayebo A.D. (1980). Role of dietary *Lactobacilli* in gastro-intestinal microbiology. *Am. J. Clin. Nutri*, 33:2448–2457.

Shahani K.M., Vakil J.R. and Kilara A. (1977). Natural antibiotic activity of *Lactobacillus acidophilus* and *bulgaricus*. II. Isolation of acidophilin from *L. acidophilus*. *Cult. Dairy Prod.* J. 12: 8–11.

Sharma V.D., Sethi M.S., Kumar A. and Rarotra J.R., (1977). Antibacterial property of *Allium sativum* Lill: *in vitro* and *in vivo* studies. Indian *Journal of Experimental Biology*. 15:466–468.

Sharon N. (1987). Bacterial lectins, cell-cell recognition and infectious disease. *FEBS Letters*. 217: 145–157.

Sharon N. and Halina L. (1993). Carbohydrates in cell recognition. *Scientific American*. 82–89.

Sharp R., O'Donnell A.G., Gilbert H.G. and Hazlewood G.P. (1992). Growth and survival of genetically manipulated *Lactobacillus plantarum* in silage. *Applied and environmental Microbiol.* 58:2517–2522.

Sharpe M.E. (1970). Cell wall and cell membrane antigens used in the classification of *Lactobacilli. Int. J. Syst. Bacteriol*, 20. 509–518.

Simm G., Chamberlain A.G. and Davies A.B. (1980). The effect of acid milk replacer and faecal coliform populations in pre-weaned calves. *Vet. Rec.* 107: 64.

Simonds J., Hensen P.A., and Lakshmanan S. (1971). Deoxyribonucleic acid hybridization among strains of *Lactobacilli. J. Bact*, 107: 382–384.

Simons P.C.M. (1990). Improvement of phosphorus availability by microbial phytase in broilers and pigs. *Brit. J. Nutr.* 64: 525–540.

Siriwan D. (1977). Effect of probiotic feeding on the performance of broiler chicks. Thesis – Poultry Science Dept. Mississippi State University.

Sissins J. (1988). In: *Practical Applications of Probiotics in Livestock Production. Probiotics Theory and Applications*. Chalcombe Publications, Wye, England : 40.

Sisson S. (1938). *The anatomy of the domestic animals* (3rd ed.). Revised by Grossman J.D., Philadelphia and London.

Skoufos A. and Fthenakis G.C. (1992). Pentosanase (Allzyme PT) supplementation of pig starter diets. *Int. Pig Vet. Soc. Proc.* 2: pp. 638.

Smith H.W. (1965a). Observations on the flora of the alimentary tract of animals and factors affecting its composition. *J. Pathol. Bact.* 89: 95–122.

Smith H.W. (1965b). The development of the flora of the alimentary tract in young animals. *J. Pathol. Bacterial.* 90: 495513.

Smith H.W. and Jones J.E.T. (1963). Observations on the Alimentary Tract and its bacterial flora in healthy and diseased pigs. *J. Path. Bact.*, 86: 387–412.

Smith H.W. (1965). Observations on the flora of the alimentary tract of animals and factors affecting its composition. *J. Path. Bact.* 89: 95–122.

Smith H.W. (1972). The antibacterial activity of nitrovin *in vitro:* the effect of this and other agents against *Clostridium welchii* in the alimentary tract of chickens. *Vet. Rec.* 90: 310–312.

Smith H.W. and Halls S. (1967). Studies on *Escherichia coli* enterotoxin J. *Path. Bact.*, 93: 531–543.

Smith R.G. (1959). The development and function of the rumen in milk-fed calves. *J. Agric. Sci.* 52: 72–78.

Smith W.H. (1971). The bacteriology of the alimentary tract of domestic animals suffering from *E coli* infection. *Annals New York Academy of Science*. 176: 110–125.

Snoeyenbos, G.H., Weinack O.M. and Smyser C.F. (1978). Protecting chicks and poults from *Salmonellae* by oral administration of "normal" gut micro-flora. *Avian Dis.* 22:273–387.

Snoeyenbos G.H., Weinack O.M. and Smyser C.F. (1979). Further studies on competitive exclusion for controlling *Salmonella* in chickens. *Avian Dis*, 23:904–914.

Soerjadi A.S., Stehman S.M., Snoeyenbos G.H., Weinack O.M. and Smyer C.F.

(1981). The influence of *Lactobacilli* on the competitive exclusion of paratyphoid *Salmonellae* in chickens. *Avian Dis.* 25: 1027–1033.

Solomons G.L. (1967) Antifoams. *Process Biochem*, 2 (10): 47–48

Solomons G.L. (1969). *Materials and Methods in Fermentation*, Academic Press, London.

Sorrels K.M. and Speck M.L. (1970). Inhibition of *Salmonella gallinarum* by culture filtrates of *Leuconostoc citrovorum*. *J. Dairy Sci*, 53:239–241.

Speck M.L. (1976). Interactions among *Lactobacillus* and man. *J.Dairy Sci*, 59:338–342.

Speck R.S., Calloway D.H. and Hadley W.K. (1970) Human fecal flora under controlled diet intake. *Am. J. Clin. Nutr.* 23: 1488–1494.

Spector W.S. (1956). *Handbook of Biological Data*. W.B. Saunders Co., Philadelphia and London.

Speth J., Greenstein M. and Maise W. (1981). The mechanism of action of avoparcin, a glycopeptide antibiotic. *Abstracts of the 81st meeting ASM Dallas:* A15.

Sprinz H. (1962). Morphological response of intestinal mucosa to enteric bacteria and its implication for sprue and Asiatic cholera. Fed. Proc. Fed. Am. Socs. exp. Biol., 21:57–64.

Steffen E.K. and Berg R.D. (1983). Relationship between cecal population levels of indigenous bacteria and translocation to the mesenteric lymph nodes. *Infect. Immun.*, 39:1252–1259.

Stephen A.M. and Cummings J.H. (1980). The microbial contribution to human faecal mass. *J. Med. Micr.* 13:45–56.

Stewart C.S., Gilmour J. and McConville M.L. (1986). Microbial interactions, manipulation and genetic engineering. In: *Agriculture – New Developments and Future Perspectives in Research on Rumen Function*. pp. 243–257, (Ed: A. Neimam-Sorensen), Commission of the European Communities.

Stewart C.S. (1977). Factors affecting the cellulolytic activity of rumen contents. *Environ. Microbiol.* 33: 497–502.

Stirling A.C. (1953). *Lactobacilli* and silage-making. *Proc. Soc. App. Bact.* 16:27–29.

Stokstad E.L.R. and Jukes T.H. (1950) The growth promoting effect of aureomycin on turkey poults. *Poultry Sci.* 29: 611–612.

Strobel H.J. and Russell J.B. (1986). Effect of pH and energy spilling on bacterial protein-synthesis by carbohydrate-limited cultures of mixed rumen bacteria. *J. Dairy Sci.* 69: 2941.

Stutz M.W. and Lawson G.C. (1984). Effects of diet and anti-microbials on growth, feed efficiency, and intestinal *Clostridium perfingens* and ideal weight of broiler chicks. *Poultry Sci.* 63: 2036–2042.

Su T.L. (1948). Micrococcin: antibacterial substance formed by strain of *Micrococcus*. *Br J. Exp. Path.* 29: 473–481.

Sutherland I.W. (1985). Biosynthesis and composition of Gram negative bacterial extracellular and wall polysaccharides *Ann. Rev. Micribiol*, 39: 243–244.

Svozil B., Danek P., Kumperecht I. and Zobac P. (1987). The efficiency of different contents of the bacterium *Streptococcus faecium M-74 in* the nutrition of calves. *Nutr. Abstr. Rev.* (Series B) 57: 615.

Swann M.M. (1969) *Report of the Joint Committee on the use of Antibiotics in Animal Husbandry and Veterinary Medicine*, London.

Swanson E.W. (1963). Effects of chlortetracycline in calf starter and milk. *J.Dairy Sci.* 46:955.

Tagg J.R., Dajani A.S. and Wannamaker L.W. (1976). Bacteriocins of Gram-positive bacteria. *Bact. Rev.* 40:722.

Takeuchi A. and Savage D. (1973) Adherence of *Lactobacilli* and yeasts to murine gastric epithelium. Paper presented at annual meeting of the *Am. Soc, Micr*, Miami, Florida USA.May 6–11.

Tamine A.Y. and Deeth H.C. (1980). *Yoghurt: Technology and Biochemistry*. Pergamon Press.

Tamine A.Y. and Robinson R.K. (1985). Microbiology of Yoghurt Starter Cultures. *Yoghurt Science and Technology*. Pergamon Press.

Tanksley, T.D., Jr. (1978). Efficacy of including a probiotic in diets for pigs weaned at 3,4 and 5 weeks of age. *Texas. Agric. Expt. Progress Rpt*: 35.

Tannock G.W. (1983). The effect of dietary and environmental stress on the gastro-intestinal microbiota. In: *Human Intestinal Micro-flora in Health and Disease* (Ed: D.J. Hentges), pp.517–539. Academic Press, New York.

Tannock G.W. and Savage D.C. (1974). Influences of dietary and environmental stress on microbial populations in the murine gastro-intestinal tract. *Inf. Immun.* 9: 591–598.

Tannock G.W. and Savage D.C. (1974). Microbial interference between indigenous yeast and *Lactobacilli* in the rodent stomach. *J. Bact.* 98:1278.

Tannock G.W. and Smith J.D.B. (1970). The micro-flora of the pig stomach and its possible relationship to ulceration on the *pars oesophagia*. *J. Comp. Path*, 80, 359–367.

Thacker P. (1988). Novel approaches to growth promotion in the pig. In: *Recent Advances in Animal Nutrition – 1988*. (Eds: W. Haresign and D.J.A. Cole). *pp. 73–84*, Butterworths, London.

Thomas R.C. and Rogers F. (1950). *Microbiology of Silage*. Ohio Agricultural Experimental Station Research Circular. No. 2. pp. 18.

Thomas R.O., Hatch R.C. (1974). Effect of *Lactobacillus acidophilus* as an additive to the feed of baby pigs. *West Virg. Agric.* For. 5: 15.

Thomlinson J.R. and Lawrence T.L.J. (1981). Dietary manipulation of gastric pH in the prophylaxis of enteric disease in weaned pigs: Some field observations. *Vet. Rec.* 109: 120–122.

Thorbecke G.J., Gordon H.A., Wostmann B., Wagner M. and Teyniers J.A. (1957). Lymphoid tissue and serum gamma globulin in young germ free chickens. *J. Inf. Dis.* 101:237–240.

Torrey J.C. and Kahn M.C. (1923). Inhibition of putrefactive spore bearing anaerobes by *Bacterium acidophilus*. *J. Inf. Dis*, 33: 482–497.

Tortuero F. (1973). Influence of the implantation of *Lactobacillus acidophilus* in chicks on the growth, feed conversion, malabsorption of fats syndrome and intestinal flora. *Poultry Sci.* 52: 197–203.

Tramer J. (1966). Inhibitory effect of *Lactobacillus acidophilus*. *Nature*. 211: 204–205.

Tsai Y., Cole L., Davis L., Lockwood S., Simmons V. and Wild W. (1985). Antiviral properties of garlic: *In vitro* effects on Influenza B, Herpes Simplex and Coxsackie viruses. *Planta Medica*. 5: 460–461.

Uden P., Colucci P.E. and Van Soest P.J. (1980). Investigation of chromium, cerium and cobalt as markers in digesta rate of passage studies. *J. Sci. Food Agric.* 31: 625–632.

Umesaki Y., Tohyama K. and Mutai M. (1982). Biosynthesis of microvillus membrane-associated glycoproteins of small intestinal epithelial cells in germ free and conventionalised mice. *J. Biochem*, 92:373–379.

Underdahl N. (1983). *Streptococcus faecium* for control of colibacillosis in pigs. *Pig News and Information* 4:(4) 435–438.

Underdahl N.R. (1983). The effect of feeding *Streptococcus faecium* upon *Escherichia coli* induced diarrhoea in gnotobiotic pigs. *Prog. Fd. Nutr. Sci*, 7: 5–12.

Vakile J.R. and Shahani K.M. (1965) Partial purification of antibacterial activity of *Lactobacillus acidophilus. Bact. Proc*, 9.

Van Houte J., Gibbons R.J. and Pulkkinen A.J. (1972) Ecology of human oral *Lactobacilli. Inf. Immun*, 6: 723–729.

Van Soest P.J, (1991). *Nutritional Ecology of the Ruminant*, 2nd Ed. Comstock Publishing Associates, Ithaca and London.

Varel V.H., Robinson I.M. and Jung H.J.G. (1987). Influence of dietary fiber on zylanolytic and cellulolytic bacteria of adult pigs. *App. Env. Micr*, 53:22–26.

Vincent J.G., Veomett R.C. and Riley R.F. (1959). Antibacterial activity associated with *Lactobacillus acidophilus. J. Bact.* 78: 477–484.

Vincent J.G., Veomett R.C. and Riley R.F. (1955). Relation of the indigenous flora of the small intestine of the rat to post-irradiation bacteria. *J. Bact*, 69: 38–44.

Vincent J.G., Veomett R.C. and Riley R.F. (1959). Antibacterial activity associated with *Lactobacillus acidophilus. J. Bact.* 78: 477–484.

Visek W.J. (1978). The mode of growth promotion by antibiotics. *J. Anim. Sci*, 46: 1447–1469.

Wadstrom T., Andersson K., Sydow M., Axelsson L., Lindgren S. and Gullman B. (1987). Surface properties of *Lactobacilli* isolated from the small intestine of pigs. *J. Appl. Bact.* 62:513–520.

Wallace K., Patterson A., McCarthy A., Paeder U., Ramsey L., MacDonald M., Haylock P. and Broad P, (1983). *The problem of lignin biodegradation*. Biochemical Soc. Symposium (Eds: C.F. Phelps and P.H. Clarke) London. UK. The Biochemical Society. 48: 87–95.

Wallace R.J. and Newbold C.J. (1992). Probiotics for ruminants. In: *Probiotics – the Scientific Basis* (Ed: R.Fuller) pp. 317–353. Chapman and Hall, London.

Watanabe K., Watanabe J., Kuramitsu S. and Maruyama H.B (1981). Comparison of the activity of ionophores with other antibacterial agents against anaerobes. *Antimicrob. Agents Chemother.* 19:519.

Watkins B.A. and Kratzer F.H. (1983). Effect of oral dosing of *Lactobacillus* strains on gut colonization and liver biotin in broiler chicks. *Poult. Sci.* 62: 2088–2094.

Watkins B.A., Miller B.F. and Neil D.H. (1982). *In vivo* inhibitory effects of

Lactobacillus acidophilus against pathogenic *Escherichia coli* in gnotobiotic-chicks. *Poult. Sci.* 61: 1298–1308.

Watkins B.A. and Miller, B.F. (1983a). Competitive gut exclusion of avian pathogens by *Lactobacillus acidophilus* in gnotobiotic chicks. *Poult. Sci.* 62:1771–1179.

Watkins B.A. and Miller B.P. (1983b). Colonization of *Lactobacillus acidophilus* in gnotobiotic chicks. *Poult. Sci.* 62: 2152–2157.

Watson D.C. (1989). *Salmonella enteritidis* – Is the feed really the source? *Feed Compounder*, Mounsey Publications, Derby, England. February 12–13.

Watson J.D. and Crick F.C. (1953). Molecular structure of nucleic acids. A structure for deoxyribose nucleic acids. *Nature*, London. 171:737–738.

Weichselbaum T.E., Hagerty J.C. and Mark H.B. Jr. (1969). A reaction rate method for ammonia and blood urea nitrogen utilizing a pentacyanonitroyloferrate catalyzed Bethelot reaction. *Anal. Chem.* 41: 848.

Weidmeier R.D., Arambel M.J. and Walters J.L. (1987). Effect of yeast culture and *Aspergillus oryzae* fermentation extract on ruminal characteristics and nutrient digestibility *J. Dairy Sci.* 70:2063–2068.

Weigmann (1899a). Versuch einer Einteilung der Milchsaurebakterien des Molkereigewerbes. *Zentbl. Bakt parasitkbe Abt*, II 5 825–831.

Weinack Olga M., Snoeyenbos G.H., Smuyser C.F. and Soerjadi A.S. (1981). Competitive exclusion of intestinal colonization of *Escherichia coli* in chicks. *Avian Dis.* 25: 696–705.

Weinack O.M., Snoeyenbos G.H. and Smyser C.F. (1979). A supplemental test system to measure competitive exclusion of *Salmonellae* by native micro-flora of the chicken gut. *Avian Dis*, 23:1019–1030.

Weinack O.M., Snoeyenbos G.H., Smyser C.F. and Soerjadi A.S. (1982). Reciprocal competitive exclusion of *Salmonella* and *Escherichia coli* by native intestinal micro-flora of the chicken and turkey. *Avian Dis*, 26:585–590.

Wenk C. (1990). Yeast cultures, *Lactobacilli* and a mixture of enzymes in diets for growing pigs and chickens under Swiss conditions: influence on the utilisation of the nutrients and energy. In: *Biotechnology in the Feed Industry*. (Ed: T.P. Lyons). pp. 315–329. Alltech Technical Publications, Nicholasville, Kentucky.

Wheater D.M, Hirch A. and Mattick A.T.R. (1951) "Lactobacillin" an antibiotic from *Lactobacillus*. *Nature*, 168:369.

Wheater D.M., Hirsch A. and Mattick A.T.R. (1952). Possible identity of Lactobacillin with hydrogen peroxide produced by *Lactobacilli*. *Nature*, 170: 623–624.

White F., Wenham G., Sharman G.A.M., Jones A.S., Rattray E.A.S. and McDonald I. (1969). Stomach function in relation to a scour syndrome in the piglet. *Br. J. Nutr.* 23: 847–857.

Whitehead H.R. (1933). A substance inhibiting bacterial growth, produced by certain strains of lactic *Streptococci*. *Biochem. J*, 27: 1793–1800.

Whitt D.D. and Savage D.C. (1981). Influence of indigenous microbiota on amount of protein and activities of alkaline phosphatase and disaccharidases in extracts of intestinal mucosa in mice. *Appl. Env. Micr*, 42:513–520.

Whitt D.D. and Savage D.C. (1988). Influence of indigenous microbiota on activities of alkaline phosphatase, phosphodiesterase I and thymidine kinase in mouse enterocytes. *Appl. Env. Micr*, 54:2405–2410.

Wiedmeier R.D., Arambel M.J and Walters J.L. (1987). Effect of yeast culture and *Aspergillus oryzae* fermentation extract on ruminal characteristics and nutrient digestibility. *J. Dairy Sci.* 70: 2063–2068.

Wiedmeier R.D. (1989). In: *9th Annual Utah Beef Cattle Field Day*, Brigham Young University, Provo, Utah.

Wiegersma N., Jansen G. and Van der Waaij D. (1982). Effect of twelve antimicrobial drugs on the colonisation resistance of the digestive tract of mice and on endogenous potentially pathogenic bacteria. *J. Hyg.* 88: 221–230.

Wilbur R.D., Catron D.V., Quinn L.Y, Speer V.C. and Hays V.W. (1960). Intestinal flora of pigs as influenced by diet and age. *J. Nutr.* 71:168–175.

Williams P.E.V. and Innes G.M. (1982). Effects of short term cold exposure on the digestion of milk replacer by young pre-ruminant calves. *Res. Vet. Sci.* 32: 383 – 386.

Williams P.E.V. and Innes G.M. (1989). Rumen probiosis: effects of addition of yeast culture (viable yeast *Saccharomyces cerevisiae* plus growth medium) on patterns of rumen fermentation. *Anim. Prod.* 48: 665 (Abstr.).

Williams P.E.V. and Newbold C.J. (1990). Rumen probiosis: The effects of novel micro-organisms on rumen fermentation and ruminant productivity. In *Recent Advances in Animal Nutrition* (Eds W. Haresign and D.J.A. Cole), pp. 211–227 London, Butterworth,

Windhozl M. (1976). The Merck Index: an encyclopedia of chemicals and drugs (9th ed.) 1285.

Wiseman J. (1990). Broiler production: market trends, meat quality and nutrition in the light of changing consumer requirements. In: *Biotechnology in the Feed Industry* pp 119–134. (Ed:T.P. Lyons). Alltech Technical Publications, Nicholasville, Kentucky.

Wolin M.J. (1981). Fermentation in the rumen and human large intestine. *Science.* 213:1463–1468.

Woods D.E., Boss J.A., Johanson Jr W.G. and Straus D.C. (1980). Role of adherence in the pathogenicity of *Pseudomonas aeruginosa* lung infections in cystic fibrosis patients. *Infect, Immun.* 30:694–699.

Woodward J. (1988). Dilemma for the meat trade over future product development. *Meat Industry. Oct*, 22–23.

Wostmann B.S., Reddy B.S., Bruckner-Kardoss E., Gordon H.A. and Singh B. (1973). In: *Germ Free Research: Biological Effect of Gnotobiotic Environments* (Ed: J.B. Heneghan), p.261. Academic Press, New York and London.

Wu J.F. (1987). The Microbiologists function in development action-specific micro-organisms. *Biotechnology in the Feed Industry.* (Ed: T.P. Lyons) Lexington, Nicholas-ville, Kentucky.

Yokota H. and Coates M.E. (1982). The uptake of nutrients from the small intestine of gnotobiotic and conventional chicks. *Br. J. Nutr.* 47: 349–356.

Young G., Krasner R.I. and Yudkofsky P.L. (1956). Interactions of oral strains of *Candida albicans* and *Lactobacilli*. *J. Bact*, 72: 525–529.

Zhu J.Q., Fowler V.R. and Fuller M. (1988). In: *Digestive Physiology in the Pig* (Eds: L. Buraczewska, S. Buraczewski, B. Pastuzewska and T.Z. Debrowska) Jablonna Institute of Animal Physiology and Nutrition, *Polish Academy of Sciences*, 180–187.

Zubrzycki L. and Spaulding E.H. (1962). Studies on the stability of the normal human faecal flora. *J. Bact.* 83: 968–974.

INDEX